Studies in Systems, Decision and Control

Volume 408

The series "Studies in Systems, Decision and Control" (SSDC) covers both new developments and advances, as well as the state of the art, in the various areas of broadly perceived systems, decision making and control–quickly, up to date and with a high quality. The intent is to cover the theory, applications, and perspectives on the state of the art and future developments relevant to systems, decision making, control, complex processes and related areas, as embedded in the fields of engineering, computer science, physics, economics, social and life sciences, as well as the paradigms and methodologies behind them. The series contains monographs, textbooks, lecture notes and edited volumes in systems, decision making and control spanning the areas of Cyber-Physical Systems, Autonomous Systems, Sensor Networks, Control Systems, Energy Systems, Automotive Systems, Biological Systems, Vehicular Networking and Connected Vehicles, Aerospace Systems, Automation, Manufacturing, Smart Grids, Nonlinear Systems, Power Systems, Robotics, Social Systems, Economic Systems and other. Of particular value to both the contributors and the readership are the short publication timeframe and the world-wide distribution and exposure which enable both a wide and rapid dissemination of research output.

Indexed by SCOPUS, DBLP, WTI Frankfurt eG, zbMATH, SCImago.

All books published in the series are submitted for consideration in Web of Science.

More information about this series at https://link.springer.com/bookseries/13304

Chao Zhai · Hai-Tao Zhang · Gaoxi Xiao

Cooperative Coverage Control of Multi-Agent Systems and its Applications

Springer

Chao Zhai
Department of Automatic Control
School of Automation
China University of Geosciences
Wuhan, China

Hai-Tao Zhang
School of Artificial Intelligence
and Automation
Huazhong University of Science
and Technology
Wuhan, Hubei, China

Gaoxi Xiao
School of Electrical and Electronic
Engineering
Nanyang Technological University
Singapore, Singapore

ISSN 2198-4182 ISSN 2198-4190 (electronic)
Studies in Systems, Decision and Control
ISBN 978-981-16-7627-7 ISBN 978-981-16-7625-3 (eBook)
https://doi.org/10.1007/978-981-16-7625-3

This Springer imprint is published by the registered company Springer Nature Singapore Pte Ltd.
The registered company address is: 152 Beach Road, #21-01/04 Gateway East, Singapore 189721, Singapore

This book is wholeheartedly dedicated to my respectable supervisors during my graduate and postdoctoral studies, as well as our group members, with whom we have worked over the years and have made it possible to reach this moment.

To my beloved families.

Preface

The rapid development of semiconductor and communication technology enables us to readily deploy, coordinate and control a large number of intelligent sensors or robots in a variety of applications. Compared to a single agent with limited sensing and actuating capabilities, a multi-agent system (MAS) is composed of plenty of intelligent agents (e.g., smart sensors, robots and unmanned systems) that communicate with others and team up to fulfill complicated collective missions, such as space exploration, formation control, border patrolling, target interception, region coverage, environment monitoring, smuggle seizing, water pollution clearance, material delivery and so on. By swarming intelligence emerged by individual agents and inter-agent interactions, cooperative control can effectively enhance the robustness, adaptability and self-organizing capabilities of MAS with low costs and high reliability. On the other hand, coverage control of MAS has been widely investigated in the past decades due to its indispensable role in wireless sensor networks, multi-robot systems and multiple unmanned systems. Essentially, it centers on how to maximize the monitoring region by sensors or enlarging the coverage rate of a given terrain to guarantee no area undetected in the region of interests. Nevertheless, the uncertainties of external environments have made it a great challenge for researchers to develop an effective cooperative control approach to fulfill the coverage tasks in given time periods, while maintaining a desired level of coverage quality.

To deal with the above issues, this monograph proposes novel theoretical formulations and practical methodologies for cooperative coverage control of MAS through the divide-and-conquer scheme. On the whole, this monograph can be divided into four parts.

The first part of this monograph presents the research background as well as the state of art literature review on cooperative coverage control approaches of MAS. In terms of multi-robot coordination, the cooperative control of MAS can be classified as sweep coverage, blanket coverage and barrier coverage. According to the final state of agents during the coverage, it can be categorized into static and dynamic coverages.

The second part focuses on the theoretical formulation and technical analysis of cooperative coverage control approaches of MAS. First of all, we develop a distributed workload partition algorithm of MAS in order to divide the coverage region into multiple sub-regions with nearly equal workload. In particular, the partition error of workload on sub-regions is estimated via the theory of input-to-state stability. Then, a decentralized sweep coverage algorithm of MAS is proposed to complete the workload on a given uncertain region by apportioning equal workload to a series of sub-stripes. As a result, each agent only needs to sweep its own sub-stripes and meanwhile partition the future stripe in a cooperative way. Moreover, we investigate how to coordinate three operations of sweeping, workload partition and communication among agents in order to enhance the coverage efficiency. Finally, a fully distributed coverage algorithm of MAS is designed with the assistance of workload memory, and the input-to-state stability is guaranteed as well.

The third part provides the practical applications of cooperative coverage control of MAS in the various fields (e.g., missile interception, intelligent traffic systems and environment monitoring). In the first case of application, a coverage-based interception approach is designed to enhance the joint probability of interception by integrating the cooperative guidance algorithm and the optimal control law for multiple interceptors of low maneuverability. The second case presents a coverage-based cooperative routing algorithms for unmanned systems to alleviate the traffic congestion in the intelligent transportation system. The third application is illustrated in the field of environment monitoring of wireless sensor networks, where a distributed coverage approach is proposed to maximize the detection probability of geohazards (e.g., landslides, debris flow and surface collapse) and forest fire for a better pre-warning mechanism.

The last part summarizes the main contributions in this monograph, and outlines some potential research directions for future work.

Wuhan, China — Chao Zhai
Wuhan, China — Hai-Tao Zhang
Singapore, Singapore — Gaoxi Xiao
July 2021

Acknowledgements

There have been many people who have walked alongside me during my academic journey. They have guided, supported, and accompanied me. I would like to, hereby, thank each of them sincerely.

First and foremost, I would like to express my deepest gratitude to my respectable supervisors Prof. Yiguang Hong at Institute of Systems Science, Chinese Academy of Science, Beijing, China, and Prof. Hai-Tao Zhang at School of Artificial Intelligence and Automation, Huazhong University of Science and Technology, Wuhan, China for their unwavering support, encouragement and constructive guidance. They, upon whose shoulders I stand, explored and paved the path before me. Without their inspiring guidance, this monograph would simply not have been possible. Besides, I would like to thank my postdoctoral supervisor Prof. Gaoxi Xiao at Nanyang Technological University, Singapore. Professor Xiao is always willing to take time to listen and usually provide insightful questions and suggestions, as well as clear instructions as feedback. His unstinting support and encouragement have driven me to strive for excellence. Many thanks are also given to Prof. Mario Di Bernardo and Dr. Time Barker at University of Bristol, Bristol, United Kingdom, Prof. Fenghua He at Harbin Institute of Technology, Harbin, China, and Prof. Michael Z. Q. Chen at Nanjing University of Science and Technology, Nanjing, China for insightful discussions on the routing algorithm in intelligent transportation systems, the cooperative interception against supersonic flight vehicles and distributed coverage control using workload memory.

Secondly, special thanks are given to many of my friends and colleagues including Dr. Guodong Shi, Dr. Xiaoli Wang, Dr. Shijian Chen, Dr. Youcheng Lou, Dr. Yutao Tang, Dr. Xiangru Xu, Dr. Jiangbo Zhang, Dr. Wenjun Song, Dr. Yanqiong Zhang, Dr. Peng Yi, Dr. Zhenhua Deng and Dr. Yinghui Wang at Institute of Systems Science, Chinese Academy of Science, Dr. Bing Ai at Huazhong University of Science and Technology for sharing a wonderful and memorable life with me. In addition, my wholehearted thanks are given to Dr. Hehong Zhang, Dr. Wenqi Du, Dr. Chaolie Ning, Dr. Yuankun Liu, Dr. Beibei Li and Dr. Min Meng at Nanyang Technological University for their generous support and understanding given in many moments of crisis over the years in Singapore. I cannot list all the names here, but you guys hold a special place in my heart.

Finally, and most importantly, my most heartfelt and forever gratitude goes to my parents Mr. Shengli Zhai and Mrs. Xiaohui Zhou and other family members including Mrs. Juanjuan Zhai, Mr. Tao Guo and Mrs. Jingjing Zhai, who have always been a constant source of support and encouragement. Thanks to my parents for putting me through the best education possible and giving me the strength to reach for the stars and chase my academic dream. I appreciate their sacrifices and unending support, and I would not have been able to get to this stage without them.

This work is supported by the Fundamental Research Funds for the Central Universities, China University of Geosciences (Wuhan) and is supported in part by the National Natural Science Foundation of China under Grant U1713203, in part by the Natural Science Foundation of Hubei Province under Grant 2019CFA005. and the Program for Core Technology Tackling Key Problems of Dongguan City under Grant 2019622101007. It is also partially supported by the Future Resilient System Project (Stages I and II) at the Singapore-ETH Centre (SEC), which is funded by the National Research Foundation of Singapore (NRF) under its Campus for Research Excellence and Technological Enterprise (CREATE) program, and also supported by Ministry of Education of Singapore under Contract MOE 2016-T2-1-119.

Contents

1 Introduction to Multi-agent Cooperative Coverage Control 1
1.1 Background 1
1.2 State of the Art 4
1.3 Outline of the Book 8
References 9

2 Distributed Control Scheme for Online Workload Partition 13
2.1 Introduction 13
2.2 Problem Statement 14
2.3 Theoretical Analysis 17
2.4 Simulation Results 24
2.5 Conclusions 25
References 25

3 Decentralized Cooperative Sweep Coverage Algorithm in Uncertain Environments 27
3.1 Introduction 27
3.2 Formulation and Coverage Algorithm 28
3.3 Technical Analysis 31
3.3.1 Key Lemmas 32
3.3.2 Main Results 38
3.4 Simulation Results 42
3.5 Conclusions 44
References 44

4 Adaptive Cooperative Coverage Algorithm with Online Learning Strategies 47
4.1 Introduction 47
4.2 Problem Formulation 48
4.3 Technical Analysis 50
4.4 Conclusions 60
References 60

5 Distributed Sweep Coverage Algorithm Using Workload Memory 63
5.1 Introduction 63
5.2 Problem Formulation 64
5.3 Key Lemmas 68
5.4 Stability Analysis 73
5.5 Numerical Simulations 78
5.6 Conclusions 79
References 80

6 Cooperative Sweep Coverage Algorithm of Discrete Time Multi-agent Systems 81
6.1 Introduction 81
6.2 Problem Formulation 82
6.3 Main Results 84
6.4 Simulations 93
6.5 Conclusions 94
References 94

7 Coverage-Based Cooperative Interception Against Supersonic Flight Vehicles 95
7.1 Introduction 95
7.2 Problem Formulation 96
7.3 Main Results 100
7.4 Numerical Simulation 107
7.5 Conclusions 109
References 109

8 Coverage-Based Cooperative Routing Algorithm for Unmanned Ground Vehicles 111
8.1 Introduction 111
8.2 Basic Idea 112
8.3 Routing Algorithm 113
8.4 Simulation Results 114
8.5 Conclusions 117
References 117

9 Cooperative Coverage Control of Wireless Sensor Networks for Environment Monitoring 119
9.1 Introduction 119
9.2 Problem Statement 120
9.3 Main Results 122
9.3.1 Coverage Region Without Obstacles 122
9.3.2 Coverage Region with Obstacles 124
9.4 Numerical Simulation 130
9.5 Conclusions 131
References 132

10 Summary and Future Work 133
10.1 Summary 133
10.2 Future Work 134

Appendix: Mathematical Concepts 137

Acronyms and Notation

Acronyms

DSCA	Decentralized sweep coverage algorithm
EDF	Event density function
HPI	Highest probability interval
ISS	Input-to-state stability
LCR	Lann-Chang-Roberts
MAP	Maximum a-posteriori probability
MAS	Multi-agent system
MMSE	Minimum mean square error
MRN	Multi-robot network
MRS	Multi-robot system
PDF	Probability distribution function
SUMO	Simulation of urban mobility
UAV	Unmanned aerial vehicle
UGV	Unmanned ground vehicle
WSN	Wireless sensor network

Notation

n	Number of agents
l_a	Width of a rectangular coverage region
l_b	Length of a rectangular coverage region
d	Diameter of actuation range for each agent
q	Number of stripes in the coverage region
$\rho(x, y)$	Workload distribution function
$\bar{\rho}$	Upper bound of workload distribution function
$\underline{\rho}$	Lower bound of workload distribution function

v	Amount of workload completed in unit time by an agent
$\lceil x \rceil$	The smallest integer larger than or equal to the real number x
R	Communication radius of each agent
$\mathbb{R}$	Set of real numbers
$\mathbb{E}_n$	Set of integers from 1 to n
n	Number of agents
N_c	Number of send-receive operations
A_k	The k-th stripe on the coverage region
$\mathbb{D}$	Sweeping rectangular region
m_i	Workload on the i-th sub-stripe
T_c	Common switching time on the stripe
T_p	Time period of one send-receive operation
T	Actual coverage time of the whole region
T^*	Optimal coverage time of the whole region
ΔT	Error between the actual and optimal coverage time
x_I	Position of the interceptor
x_T	Position of the target
v_I	Velocity of the interceptor
v_T	Velocity of the target

Chapter 1
Introduction to Multi-agent Cooperative Coverage Control

1.1 Background

In the past hundred years, scientists mainly revolve around the single object or simple systems. As for large-scale complex systems, the basic solution is to decompose the complex system into multiple simple subsystems. The above research philosophy largely ignores the interdependence among various subsystems and thus fails to explain some peculiar phenomena, such as flock of birds, fish schools in the sea, herd behaviors, and wild geese flying in a V formation, as shown in Fig. 1.1. As a result, new research paradigms emerge by inquiring into the complex system as a whole instead of investigating each component separately. More and more researchers have been attracted by this new field, which is regarded as "Science in the 21st Century" [1–3].

As an important theoretical model of complex systems, multi-agent system (MAS) is generally composed of smart agents and external environments that can affect the evolution of agents [4–7]. Each agent is able to sense the surrounding environment and obtain the information of interest, and it can also communicate with neighboring agents to share the information. Based on the information and control rules, agents can work together to accomplish complicated missions, which will further cause the change of external environments. In MAS, each agent can implement the control command independently according to the specified rule, and the whole system displays collective behaviors. When the scale of MAS becomes rather large, the centralized control approaches are vulnerable to external attacks or unknown disturbances. In comparisons, decentralized control scheme is more attractive as it can provide flexible and robust solutions to uncertainties from external environments or intrinsic dynamics of agents.

In the past decades, researchers in the fields of computer science, social science and systems science have made great progress in designing decentralized control strategies and applying them to engineering problems [8–13]. Most scholars in computer science, biology and statistic physics prefer to investigate collective behaviors of animals with the assistance of numerical simulations. In [8], the first computer

C. Zhai et al., *Cooperative Coverage Control of Multi-Agent Systems and its Applications*, Studies in Systems, Decision and Control 408,
https://doi.org/10.1007/978-981-16-7625-3_1

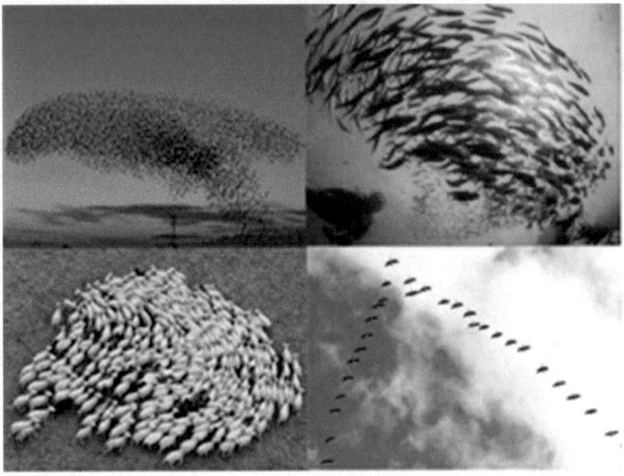

Fig. 1.1 Coordination behaviours of animal groups in nature

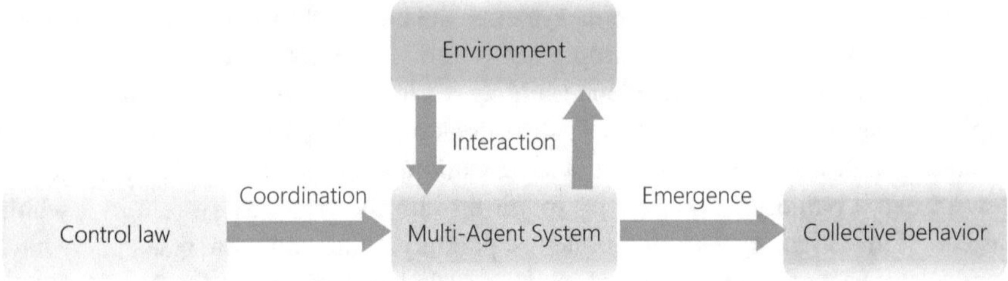

Fig. 1.2 Interaction between MAS and external environments

code is created to model the flocking behaviors of birds, and three basic rules are proposed to explain how collective behaviors emerge in biology. These three rules include: (1) Aggregation for getting close to their neighbors, (2) collision avoidance for preventing the collision, and (3) alignment for catching up with their neighbors. In [9], a simple yet effective model is developed to reveal the consensus and phase transition of mobile particles. In comparisons, researchers in mathematics and control science tend to conduct theoretical analysis on the convergence, robustness and stability of dynamical systems that capture the behavior of agents. Figure 1.2 demonstrates how the collective phenomena emerge through the interaction between agent dynamics and external environment from a perspective of control science. Here, the key challenge is to design the cooperative controller for each agent so that MAS displays the desired collective behaviors.

In recent years, much significant progress have been made in the field of multi-agent cooperative control, which includes multi-agent output regulation, multi-agent set coordination and multi-agent cooperative coverage. Therein, multi-agent output regulation is an important research topic in modern control theory due to its research background and practical values in applications [14]. Essentially, the goal of distributed output regulation is to achieve the asymptotic tracking and disturbance rejection by designing distributed feedback controller, and the reference signal and disturbance come from exterior systems. To solve this problem, the exterior system is treated as the leader of MAS. A practical example related to multi-agent output regulation is presented as follows: How to estimate the position of dynamic target using sensor networks, where followers exchange measurement information for the consensus with the leader. In such problems, states of exterior systems are only available to some agents, and distributed estimation and distributed internal model principle are employed to accomplish the distributed output regulation. In the distributed estimation, the exterior system is regarded as a “dynamic leader”. Then the distributed estimator or observer can be constructed according to system structures for estimating the states of leader, which helps to achieve the desired goal of MAS. Compared to internal model principle, the controller based on distributed estimation is relatively simple and applied to simple systems, despite the lack of a unified theoretical formulation. On the basis of internal model principle, [15] provides a more systemic and robust approach for output regulation problem, and [16] presents a sufficient and necessary condition for the solution of output regulation problem under fixed topology. In addition, [17] considers the output regulation problem of heterogeneous nonlinear MAS with uncertainties and fixed special topology.

In multi-agent cooperative control, the collective goal is possibly related to one or several sets. For example, MAS (e.g., a robot team) has a target set that each agent converges to. Thus, it is of great significance to control a MAS for entering the target set. Moreover, each agent only optimizes its own index during the set optimization while the global objective is to achieve the optimization of whole system. The information exchanging among agents is usually time-varying, which makes it challenging to deal with distributed set optimization of MAS with time-varying topology. With the assumption of convexity, [18] investigates the MAS with static target set and designs a simple nonlinear control rule that drives the MAS entering a target set and obtains the connectivity condition. In addition, many practical problems can be formulated with the assistance of set coordination and multi-leader cooperation. Reference [19] develops a containment control scheme of MAS with undirected graph that enables followers to move into a polyhedron composed of multiple mobile leaders. Distributed optimization algorithms are also proposed and they largely include gradient method, duality decomposition method and intelligent algorithms. Reference [20] presents a distributed algorithm for special optimization problems, and [21] succeeds in solving a decoupling optimization problem using sub-gradient method and consensus algorithms. With the assumption of time-varying topology, [22] estimates the convergence rate of the sum of convex objective functions. Reference [23] discusses the consensus problem in convex optimization with switching topology and constraints. In duality decomposition method, each agent is able to update its own local states and converge to a common value for global optimization [24]. In addition, some progress has been made in heuristic intelligent algorithms [25].

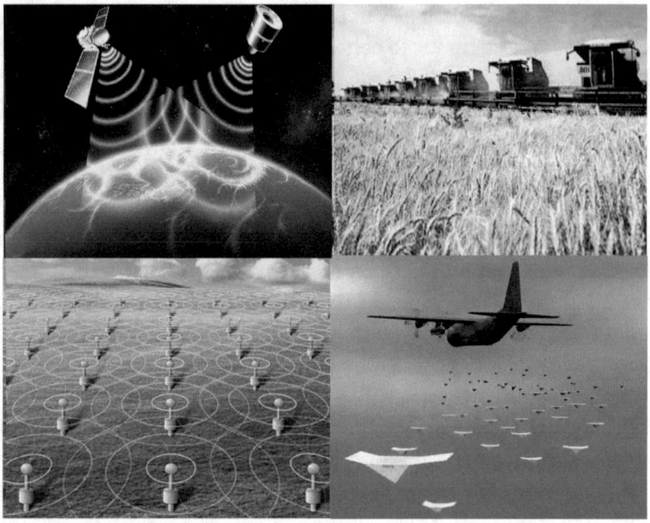

Fig. 1.3 Practical applications of multi-agent cooperative coverage control

1.2 State of the Art

Cooperative coverage control of MAS has been deeply investigated by scholars from different perspectives and found wide applications in various fields (see Fig. 1.3). The research approaches can be summarized as cellular decomposition method, formation control method, error function method, swarm intelligence method and game theoretic method (see Fig. 1.4). In the cellular decomposition method, the whole coverage region is divided into multiple sub-regions, and each agent is only responsible to cover its own sub-region by communicating with its neighbors, which allows to simplify the coverage mission via task assignment and has been widely adopted. In formation control method, cooperative control algorithms are designed so that the MAS converges to a predefined formation. In this way, the coverage problem is converted into the route planning problem of MAS. The basic idea of error function is to construct an error-based energy function and drive each agent towards the area of low coverage level in order to achieve the desired coverage level. The control algorithm of agents is normally designed by adopting gradient descent method. Inspired by collective behaviors in biology, each agent leaves the traces (e.g., pheromone) after covering an area, which indicates that this area has been covered to prevent the repetitive coverage. As a result, the traces act as the bridge or medium that contributes to the interaction between agents and external environment. This method allows to cover the complicated region (e.g., non-convex region), although it is difficult to do theoretical analysis rigorously. If each agent is treated as a rational individual, it is also feasible to investigate the cooperative coverage problem in the framework of game theory. For instance, potential game can be employed, where the objective

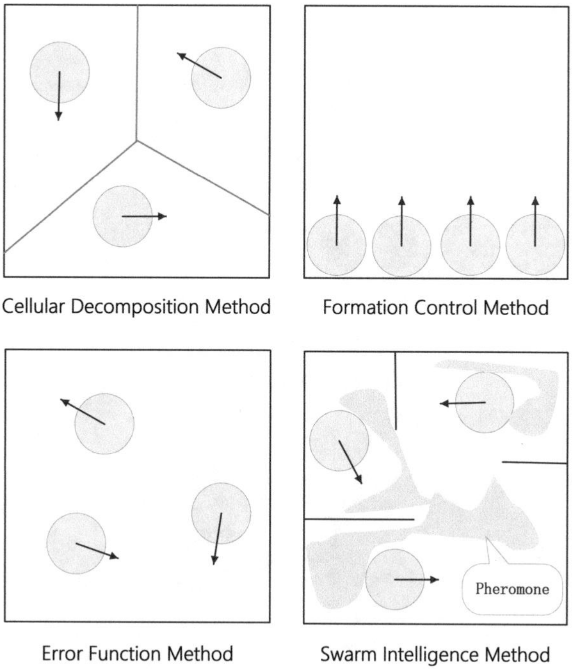

Fig. 1.4 Main methods of multi-agent coverage

function in the coverage problem is regarded as a global potential function, and the local utility function is designed for each agent. The global coverage goal can be achieved only if each agent constantly optimizes its own utility.

Cooperative coverage problem of MAS can be classified from various perspectives. In terms of mathematical tools, there are deterministic coverage and stochastic coverage. In multi-robot systems, coverage problems include blanket coverage, barrier coverage and sweep coverage [26] (see Fig. 1.5). Blanket coverage aims to deploy multiple agents in a given region in order to maximize the probability of identifying the detrimental events. Barrier coverage cares about how to form a barrier in order to protect the target inside while maximizing the detection probability of invaders crossing the barrier. To put it simply, sweep coverage can be regarded as a mobile barrier, and its goal is to sweep or monitor a coverage region by organizing MAS in some way [27, 28]. In the coverage process, the status of agents is different, which allows to classify the coverage problem into static coverage and dynamic coverage. As for the static coverage, agents are deployed at fixed locations so that each point in the region can be monitored all the time. It is necessary to possess enough agents for the static coverage of a given region, which could be difficult or even impossible because of unknown environments [29]. If there are not enough agents for static coverage, MAS can move to visit each point in the region at a certain frequency, which is called "dynamic coverage". In what follows, multi-agent coverage problem

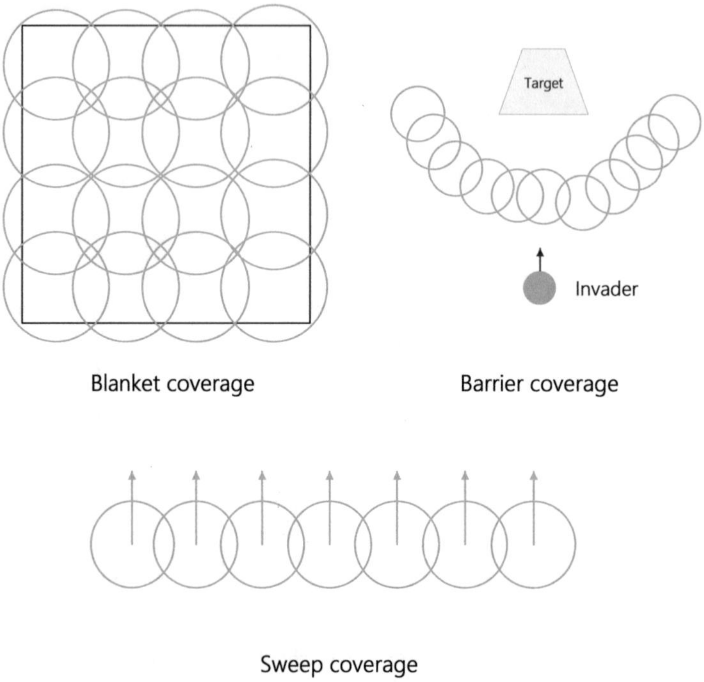

Fig. 1.5 Three coverage modes in multi-robot systems

is classified into region coverage, boundary coverage and target coverage according to coverage objects.

- **Region Coverage** One important strategy for multi-agent region coverage is to divide the whole coverage region into multiple sub-regions, and each agent only covers its own sub-region. This idea of divide-and-conquer has been widely accepted by researchers to develop coverage control approaches, such as Voronoi partition and equal workload partition. Inspired by Lloyd algorithm [30], an iterative algorithm is designed based on Voronoi partition and Parallel Axis Theorem. First of all, the coverage region is partitioned according to Voronoi diagram and initial locations of agents. The the center of mass for each sub-region is computed and each agent moves towards the center of mass in its own sub-region. After that, the coverage region is partitioned according to new locations of agents. The above process does not pause until the algorithm converges. This approach enables to accomplish the optimal coverage task in a distributed manner despite of high computation complexity. Reference [31] presents an adaptive distributed algorithm to deploy wireless sensors and optimize the objective functions by learning the probability density function of environments in real time. A consensus rule is also designed in order to speed up the learning rate of unknown parameters. In the end, the wireless sensor network is guaranteed to form a sub-optimal configuration. If the environment information is rich enough (i.e., persistent excitation condition holds), the optimal configuration is ensured. Based on the work presented in [31, 32] applies the adaptive control approach to distributed coverage of a given region and achieves the approximately optimal deployment of MAS on the condition of unknown density function. It is proved that location optimization problem of nonholonomic multi-robot system (MRS) is resolved for uncertain environment parameters by designing the adaptive coverage control algorithm. As an important type of region coverage, cooperative coverage control of WSN aims to deal with two key problems. One is about the optimal deployment of WSN by optimizing the location of static sensors and its sensing radius. The other is the redeployment problem of mobile sensors. One well-known example on the optimal deployment of WSN is the so-called "gallery problem", where some monitors are required to install in a non-convex polygon to ensure that each point in the polygon can be covered by one monitor at least. The objective is to accomplish the above mission with the least number of monitors. Reference [33] investigates this problem with the assistance of Voronoi partition and it is demonstrated that the optimal location of monitors is exactly the center of mass for each sub-region. Reference [34] considers the static region coverage from a perspective of algebraic topology and provides a sufficient condition in the sense of topology. In addition, the redeployment problem comes from many practical applications. The task is to design the control strategy for fixed number of sensors for optimizing the monitoring performance in dynamic environments. Sweep coverage deals with the problem of how to coordinate agents for completing the sweeping task in a given region [27]. Reference [35] designs a sweep coverage algorithm for several mobile robots with finite angles of view in order to visit every point of the coverage region. In light of market mechanism, [36] develops a heuristic algorithm for multi-robot network (MRN), which enables to cooperatively sweep a region with obstacles. Reference [37] proposes a distributed coverage algorithm to cover a rectangular region with a team of square robots. To detect random events in the environment, [38] presents a deployment algorithm of sensors to maximize the joint detection probability of WSN. In nature, some insects can communicate with each other with the assistance of chemical particles (e.g., pheromone). Based on such cooperation mechanism, [39] designs a novel ant robot that leaves the "chemical particle" on its trace and decides the moving direction according to the concentration of chemical particles. This bio-inspired algorithm is able to adapt to the ever-changing environment, external unknown disturbances and even malfunctions of ant robots well. To achieve the blanket coverage, [40] adopts the formation control method to cover a given region between two line segments by deploying MAS into a triangle grid. In addition, [41] propose an awareness coverage algorithm to solve the "effective coverage" problem of MAS in unknown environments. The k-coverage of MAS requires that each point of a given region can be covered by k agents at least. To save the energy, [42] constructs a k-coverage model of possibly sleepy sensors and estimates the critical density of sensors for achieving the k-coverage with the probability of 1. With the fan-shaped sensing radius, [43] provides a sufficient condition of MAS to guarantee the k-coverage with the probability of 1.

- **Boundary Coverage** Compared with region coverage, boundary coverage aims to cover the boundary of concern in a given region [44], and it mainly includes static boundary coverage and dynamic boundary coverage. For example, a group of UAVs are required to put out fires in a large region, and it is necessary to determine the fire boundary and cover it, which enables to prevent the expansion of fires effectively. As the fires expand with time, the boundary changes as well. This is obviously a problem of dynamic boundary coverage. Barrier coverage looks at how to arrange MAS to form a barrier and maximize the probability of identifying the invaders crossing the barrier. Reference [45] investigates an interesting problem of barrier coverage with the aid of computational geometry and Voronoi diagram. With the approach of formation control, [46] designs a distributed algorithm of MAS for the isometric barrier coverage of a line segment, which allows to identify targets that cross the line segment.
- **Target Coverage** The focus of target coverage is to cover the dynamic targets with uncertainties. As a result, the coverage region is not specified beforehand but evolves with the dynamic target and uncertain environments. For this reason, it is necessary to estimate the region where targets are expected to appear and then enclose the estimated region. In target coverage, the above region or its boundary has to be fully covered or under surveillance. Note that each agent is equipped with the sensors and actuators. If the target is located in the sensing region of one agent, the target can be monitored or captured by the MAS. As for mobile targets, agents have to cooperate with each other for detecting and estimating the target state in order to track and confine it. Normally, the sensing or actuation region is described by a disc, whose center is occupied by the agent. In the chase problem, the goal is to cover or contain the escaping agent with the sensing region of MAS. Reference [47] proposes a coordination approach of MAS to track and confine a mobile target. Reference [48] discusses the multi-agent chase and escape problem by simulations and optimizes the chase scheme with task assignment. Moreover, [49] considers the problem of how to track and cover a mobile target in finite time, and each agent has the limited actuation capability. Since sensors on the agent can only detect the partial states of the target (e.g., velocity, position), a cooperative observer is designed to reconstruct the state of mobile target, which allows to develop a distributed controller for tracking the mobile target with uncertainties. Moreover, the bound of estimation error is obtained for the above coverage control algorithm.

1.3 Outline of the Book

This remainder of this book is organized as follows: Chap. 2 proposes a distributed online algorithm for workload partition in uncertain environments. Based on the partition law, a decentralized sweep coverage algorithm of MAS is designed to complete the workload in uncertain regions in Chap. 3. To deal with environmental uncertainties, adaptive sweep coverage algorithm is developed with the assistance of online learning strategies in Chap. 4. By using workload memory, a fully distributed algorithm is proposed to achieve the sweep coverage in Chap. 5. Besides continuous time coverage algorithm, a discrete time control algorithm is designed for MAS in uncertain environments in Chap. 6. Afterwards, the applications of cooperative coverage algorithms are discussed in Chaps. 7, 8 and 9. Specifically, Chap. 7 considers the cooperative interception problem against supersonic flight vehicles and presents a coverage-based cooperative guidance algorithm to deploy multiple interceptors for maximizing the joint interception probability. In Chap. 8, a coverage-based routing algorithm is developed for unmanned ground vehicles to relieve the congestion level of intelligent transportation systems. Chapter 9 focuses on cooperative coverage control of wireless sensor networks for environment monitoring, which has applications in geohazards monitoring and early warning. Finally, Chap. 10 summarizes the whole book and discusses the potential research topics in the future.

References

1. Waldrop, M.: Complexity: The Emerging Science at the Edge of Order and Chaos. Simon and Schuster, New York (1992)
2. Couzin, I.D., Krause, J., Franks, N., Levin, S.: Effective leadership and decision-making in animal groups on the move. Nature **433**(3), 513–516 (2005)
3. Anderson, P.: More is different. Science **177**(4), 393–396 (1972)
4. Martínez, S., Cortés, J., Bullo, F.: Motion coordination with distributed information. IEEE Control Mag. **27**(4), 75–88 (2007)
5. Ren, W., Beard, R.W.: Distributed Consensus in Multi-vehicle Cooperative Control. Communications and Control Engineering Series. Springer, London (2008)
6. Lynch, N.A.: Distributed Algorithms. Morgan Kaufmann, San Francisco (1997)
7. Lin, P., Jia, Y., Li, L.: Distributed robust H_∞ consensus control in directed networks of agents with time-delay. Syst. Control Lett. **57**(8), 643–653 (2008)
8. Reynolds, C.W.: Flocks, herds, and schools: a distributed behavioral model. Comput. Graph. **21**(1), 25–34 (1987)
9. Vicsek, T., Cziroók, A., Ben-Jacob, E., Cohen, O., Shochet, I.: Novel type of phase transition in a system of self-deriven particles. Phys. Rev. Lett. **75**(6), 1226–1229 (1995)
10. Howard, A., Parker, L.E., Sukhatme, G.S.: Experiments with a large heterogeneous mobile robot team: exploration, mapping, deployment and detection. Int. J. Robot. Res. **25**(3), 431–447 (2006)
11. Olfati-Saber, R., Fax, J.A., Murray, R.M.: Consensus and cooperation in networked multi-agent systems. Proc. IEEE **95**(1), 215–233 (2007)
12. Lin, P., Jia, Y.: Consensus of second-order discrete-time multi-agent systems with nonuniform time-delays and dynamically changing topologies. Automatica **45**(9), 2154–2158 (2009)
13. Hong, Y., Chen, G., Bushnell, L.: Distributed observers design for leader-following control of multi-agent networks. Automatica **44**(5), 846–850 (2008)
14. Huang, J.: Nonlinear Output Regulation: Theory and Applications. SIAM, Phildelphia (2004)
15. Francis, B.A., Wonham, W.M.: The internal model principle of control theory. Automatica **12**(4), 457–465 (1976)
16. Hong, Y., Wang, X., Jiang, Z.: Multi-agent coordination with general linear models: a distributed output regulation approach. In: Proceedings of IEEE International Conference on Control and Automation, pp. 137–142, June 9-11. Xiamen, China (2010)
17. Wang, X., Hong, Y., Huang, J., Jiang, Z.: A distributed control approach to a robust output regulation problem for linear systems. IEEE Trans. Autom. Control **55**(12), 2891–2896 (2010)

18. Shi, G., Hong, Y.: Global target aggregation and state agreement of nonlinear multi-agent systems with switching topologies. Automatica **45**(5), 1165–1175 (2009)
19. Jin, M., Ferrari-Trecate, G., Egerstedt, M., Buffa, A.: Containment control in mobile networks. IEEE Trans. Autom. Control **53**(8), 1972–1975 (2008)
20. Johansson, B., Rabi, M., Johansson, M.: A randomized incremental subgradient method for distributed optimization in networked systems. SIAM J. Optim. **20**(3), 1157–1170 (2009)
21. B. Johansson, T. Keviczky, M. Johansson, and K. Johansson, Subgradient methods and consensus algorithms for solving convex optimization problems. In: Proceedings of the IEEE Conference on Decision and Control, pp. 4185–4190. Cancun, Mexico (2008)
22. Nedić, A., Ozdaglar, A.: Distributed subgradient methods for multi-agent optimization. IEEE Trans. Autom. Control **54**(1), 48–51 (2009)
23. Nedić, A., Ozdaglar, A., Parrilo, P.A.: Constrained Consensus and Optimization in Multi-Agent Networks. IEEE Trans. Autom. Control **55**(4), 922–938 (2010)
24. Xiao, L., Johansson, M., Byod, S.: Simultaneous routing and resource allocation via dual decomposition. IEEE Trans. Commun. **52**(7), 1136–1144 (2004)
25. Alba, E., Troya, J.: A survey of parallel distributed genetic algorithms. Complexity **4**(4), 31–52 (1999)
26. Gage, D.W.: Command control for many-robot systems. In: Proceedings of the Annual AUVS Technical Symposium, pp. 22–24. Huntsville, Alabama (1992)
27. Zhai, C., Hong, Y.: Decentralized sweep coverage algorithm for multi-agent systems with workload uncertainties. Automatica **49**(7), 2154–2159 (2013)
28. Zhai, C., Xiao, G., Chen, M.Z.: Distributed sweep coverage algorithm of multi-agent systems using workload memory. Syst. Control Lett. **124**, 75–82 (2019)
29. O'Rourke, J.: Art Gallery Theorems and Algorithms. Oxford University Press, New York (1987)
30. Du, Q., Faber, V., Gunzburger, M.: Centroidal Voronoi tessellations: applications and algorithms. SIAM Rev. **41**(4), 637–676 (1999)
31. Schwager, M., Slotine, J., Rus, D.: Decentralized adaptive coverage control for networked robots. Int. J. Robot. Res. **28**(3), 357–375 (2009)
32. Luna, J., Fierro, R., Abdallah, C., Wood, J.: An adaptive coverage control algorithm for deployment of nonholonomic mobile sensors. In: Proceedings of IEEE Conference on Decision and Control, pp. 1250–1256. Atlanta (2010)
33. Drezner, Z.: Facility Location: A Survey of Applications and Methods ser, Springer Series in Operations Research. Springer, New York (1995)
34. de Silva, V., Ghrist, R.: Coverage in sensor networks via persistent homology. Algebraic Geom. Topol. **7**(2), 339–358 (2007)
35. Choset, H.: Coverage for robotics – a survey of recent results. Ann. Math. Artif. Intell. **31**(1), 113–126 (2001)
36. Min, T., Yin, H.: A decentralized approach for cooperative sweeping by multiple mobile robots. In: Proceedings of the IEEE/RSJ International Conference on Intelligent Robots and Systems, pp. 380–385. Victoria, B.C., Canada (1998)
37. Butler, Z., Rizzi, A., Hollis, R.: Complete distributed coverage of rectilinear environments. In: Proceedings of the Workshop on the Algorithmic Foundations of Robotics (2000)
38. Cassandras, C.G., Li, W.: Sensor networks and cooperative control. Eur. J. Control. **11**(4–5), 436–463 (2005)
39. Wagner, I., Lindenbaum, M., Bruckstein, A.: Distributed covering by ant-robots using evaporating traces. IEEE Trans. Robot. Autom. **15**(5), 918–933 (1999)
40. Cheng, T.M., Savkin, A.V.: Decentralized control of mobile sensor networks for triangular blanket coverage. In: American Control Conference, pp. 2903–2908. Baltimore (2010)
41. Hussein, I.I., Stipanović, D.M.: Effective coverage control for mobile sensor networks with guaranteed collision avoidance. IEEE Trans. Control Syst. Technol. **15**(4), 642–657 (2007)
42. Kumar, S., Lai, T.H., Balogh, J.: On *k*-coverage in a mostly sleeping sensor networks. In: Proceedings of the 10th International Conference on Mobile Computing and Networking, pp. 144–158. Philadelphia (2004)

43. Shi, G., Hong, Y.: Region coverage for planar sensor network via sensing sectors. In: Proceedings of IFAC World Congress, pp. 4156–4161. Seoul, Korea (2008)
44. Song, C., Fan, Y.: Coverage control for mobile sensor networks with limited communication ranges on a circle. Automatica **92**, 155–161 (2018)
45. Meguerdichian, S., Koushanfar, F., Potkonjak, M., et al.: Worst and best-case coverage in sensor networks. IEEE Trans. Mobile Comput. **4**(1), 84–92 (2005)
46. Cheng, T.M., Savkin, A.V.: A distributed self-deployment algorithm for the coverage of mobile wireless sensor networks. IEEE Commun. Lett. **13**(11), 877–879 (2009)
47. Schumacher, C.: Ground moving target engagement by cooperative UAVs. In: Proceedings of American Control Conference, pp. 4502–4505, Portland (2005)
48. Oh, S., Schenato, L., Chen, P., Sastry, S.: Tracking and coordination of multiple agents using sensor networks: system design, algorithms and experiments. Proc. IEEE **95**(1), 234–254 (2007)
49. Wang, X., Hong, Y., Jiang, Z.: Coverage tracking of a moving target by a group of mobile agents. In: Proceedings of Asian Control Conference, pp. 332–337. Hong Kong, China (2009)

Chapter 2
Distributed Control Scheme for Online Workload Partition

2.1 Introduction

In recent years, the ubiquitous phenomena of cooperative behaviors among individuals in nature and engineering have attracted more and more scholars from various fields. Evidently, it is of great challenge for researchers to explore the simple mechanism underlying collective phenomena. Promisingly, multi-agent systems have provided a powerful mathematical framework or a suitable theoretical model to characterize the cooperative behaviors in complex systems or networks.

In practice, distributed design with advantages such as low cost, reliability, and flexibility provides a feasible way to deploy a large number of networked agents over a region of interest to achieve desired collective tasks. As we know, agents are usually equipped with sensors or actuators. Since a single agent may be difficult to complete the task due to its limited capacities, a group of agents (maybe viewed as a sensor network) are usually teamed up to complete the tasks by communicating and coordinating their actions through the interconnection network. The coordination algorithms for multiple agents have been deeply investigated in the literature to deal with multi-agent consensus, target tracking, and region coverage [1–4].

Coverage problems of multiple agents have drawn much attention to researchers recently [5, 6]. It is related to many challenging tasks including search and rescue, surveillance and environmental monitoring, which keeps under investigation, especially for the coordination of autonomous robots based on static or mobile networks [7–9]. Generally, it is not easy to avoid the overlap or repetition of work done by individual agents in multi-agent networks, especially when performing the coverage tasks in a region with irregular geometric shapes or uncertain distribution density. To deal with the problem, an approach based on workload partition is a way for optimization of collective behaviors.

It is well known that Voronoi partition [10] and equal workload partition [11] were widely adopted in the coverage design of MAS. For instance, the location cost function was introduced as an index to optimize sensor location in [12]. A low value of it corresponds to good coverage and a high value of it corresponds to poor

C. Zhai et al., *Cooperative Coverage Control of Multi-Agent Systems and its Applications*, Studies in Systems, Decision and Control 408,
https://doi.org/10.1007/978-981-16-7625-3_2

coverage. Moreover, in light of Parallel Axis Theorem and Lloyd algorithm [13], the index has well-defined minima corresponding to optimal coverage configurations as the centroid of Voronoi cells in the partition [10]. Furthermore, a decentralized algorithm was proposed to achieve equal workload partition with the aid of power diagrams, and some extensions of the algorithm were also taken into account [11]. Additionally, to deal with the parameter uncertainty in the covered environment, an adaptive control strategy was investigated for nonholonomic mobile sensors in [14].

The objective of this work is to design a distributed algorithm of equal workload partition for mobile agents in the environment with uncertainties. If each agent knows the environment information in advance, it can follow an optimized path to complete the coverage task. Nevertheless, limited sensing range of agents and unknown environment influences cooperative behaviors between mobile agents, and the online estimation algorithm will be proposed in order to partition the whole region effectively. Moreover, uncertain distribution density inevitably results in relatively low accuracy of workload partitions, which will be quantitatively estimated in our research.

This chapter is organized as follows. In Sect. 2.2, the formulation of finite-time workload partition for dynamical multi-agent coverage is presented. Main results on the workload partition and target estimation are obtained along with simulation examples in Sects. 2.3 and 2.4. Finally, concluding remarks and future research are given in Sect. 2.5.

2.2 Problem Statement

In this section, we formulate the problem of online workload partition as follows. For a group of agents with the aim of dynamically sweeping a region with unknown workload distribution, it is advisable to partition the whole region into sub-regions with equal workload, in order to reduce repetitive coverage and minimize the time of covering the whole region. Here for simplicity, we will not study the actuation of sweep coverage. Instead, we focus on the partition of the region, and we regard the agents as the partition points. In other words, we focus on designing a decentralized law to average the workload in the region so that multi-agent system can partition the region efficiently.

In our problem, we consider a rectangular region D with length l_a and width l_b for convenience (see Fig. 2.1), and $\rho(x, y) > 0$ describes the workload distribution in the region (that is, it is a continuous distribution density function on D). The distribution is not known beforehand. Instead, it can only be measured online during the partition procedure.

Without loss of generality, let θ denote the angle between the positive x-axis and the short edge L_b in the counter-clockwise direction as shown in Fig. 2.1. Assume the point $\mu = (\mu_x, \mu_y)^T$ is located at the long edge L_a, and the segment L_μ in the rectangular D that passes μ and is parallel to the short edge is denoted by,

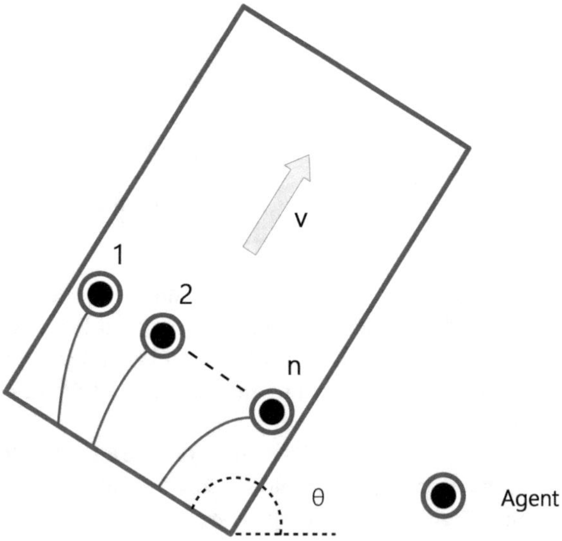

Fig. 2.1 Workload partition of multiple mobile agents on the rectangular region

$$L_\mu = \{q \in R^2 | q = \mu + l\sigma(\theta), 0 \leq l \leq l_b\},$$

where $\sigma(\theta) = [\cos\theta, \sin\theta]^T$. Let the set $E_n = \{1, 2, ..., n\}$, where n is the sum of agents, and the position of agent $i \in E_n$ is denoted by $p_i = (p_{ix}, p_{iy})^T$, and two (time-varying) endpoints of $L_\mu(t)$ denoted by $p_0(t)$ and $p_{n+1}(t)$.

The distributed method of workload partition for the multi-agent system will be elaborated in what follows. We suppose that n agents are initially lined up on the line segment L_μ (see Fig. 2.1), and each agent can obtain local information as follows. At each time, every agent can only detect the positions of its nearest neighbors on the segment and environmental information (distribution density of workload) on the corresponding interval. In addition, the agents nearest to endpoints can yield the environmental information on the subsegment between them and endpoints. It is natural to present the following assumption: n agents are lined up and connected to their neighbors next to them for all $t \geq 0$.

By regarding the position of each agent on L_μ as the boundary point of the subsegment, we can design a distributed control law in order to partition the segment into subsegments with workload as equal as possible. On the other hand, the agents have to move to dynamically partition the considered region. Assume n agents on L_μ move with the common velocity v from L_b to the other end. Therefore, to complete the dynamic region coverage, these agents move along some direction and meanwhile they average the workload on the moving segment by the proposed workload partition law. Clearly, the rectangular region D is divided into $n + 1$ subregions by the trajectories of n agents. It is obvious that a slower v will lead to the more accurate equal-workload partition when dealing online with unknown distribution. In other words, for a given workload distribution, its change along the longer edge L_a can be described by a change rate once v is given. Namely, v can be used to measure how large the distribution change is along L_a. Roughly speaking, the change rate of the distribution density can be written as

$$\begin{aligned}\frac{d}{dt}\rho(x(t), y(t)) &= \rho'(x, y) \cdot v(t) \\ &= \lim_{r \to 0} \frac{\rho(x + r \sin\theta, y - r\cos\theta) - \rho(x, y)}{r} \cdot v(t)\end{aligned}$$

Note that $\rho'(x, y)$ only depends on the space. It is fixed once the region and workload distribution are given. So the partition convergence is directly related to $v(t)$ when the real-time distribution change appears as a disturbance during the dynamic partition procedure. In this sense, v becomes a measurement of the change rate of the distribution.

To study how much the change of distribution influence the partition along the time, we employ a well-known concept called input-to-state stability (ISS) [15]. Before introducing ISS, we give the following definitions: A function $\gamma : \mathbb{R}_+ \to \mathbb{R}_+$ is said to be a K-function if it is strictly increasing and continuous with $\gamma(0) = 0$, and it is a K_∞-function if it is a K-function and $\gamma(s) \to \infty$ as $s \to \infty$. A function $\beta : \mathbb{R}_+ \times \mathbb{R}_+ \to \mathbb{R}_+$ is a KL-function if, for each fixed $t \geq 0$, the function $s \mapsto \beta(s, t)$ is a K-function, and for each fixed $s \geq 0$, the function $t \mapsto \beta(s, t)$ is continuous and decreases to zero as $t \to \infty$.

Definition 2.1 Consider a system

$$\dot{z} = f(z, u), \quad f(0, 0) = 0, \quad z \in \mathbb{R}^{n_z}, \ u \in \mathbb{R}^{n_u}, \tag{2.1}$$

where f is smooth with respect to (z, u) and the input u is measurable and locally essentially bounded. System (2.1) is said to be input-to-state stable (ISS) with u as the input if, for initial state $z(0)$ and measurable and locally essentially bounded input u, we have

$$||z(t)|| \leq \beta(||z(0)||, t) + \gamma(||u||_{[a,b]}) \tag{2.2}$$

where γ is a K-function and β is a KL-function with $\|\cdot\|$ standing for the Euclidean norm and $||u||_{[a,b]} = \text{ess.}\sup_{\tau \in [a,b]} ||u(\tau)||$.

Then, we present the mathematical expression of each agent and the corresponding control algorithm. The basis model of agent i is:

$$\dot{p}_i = \sigma(\theta) u_i, \quad i = 1, ..., n, \tag{2.3}$$

where u_i denotes the cooperative control input for agent i. Let $\lambda_i = \| p_i - \mu \|$ denote the distance between p_i and μ. Then we propose a decentralized algorithm to average the workload on the line segment. To be specific, suppose that n agents with distinct initial positions are placed on L_μ, which is a time-varying segment in D. For agent i with dynamics (2.3), we employ the following partition control law

$$u_i = \dot{\lambda}_i = m_i - m_{i+1}, \tag{2.4}$$

where

$$m_i(t) = \int_{s_{i-1,i}} \rho(x, y)ds,$$

and $s_{i-1,i}$ denotes the subsegment between p_{i-1} and p_i.

2.3 Theoretical Analysis

In this section, we give the technical analysis on the online workload partition of multi-agent system. To deal with the partition problem, we first give a lemma to discuss the convergence of the distributed control algorithm for the case when agents do not have prior information on workload distribution.

Lemma 2.1 *If n agents on L_μ in the form of (2.3) under partition control law (2.4) have distinct initial positions, then L_μ can be divided into $n + 1$ subsegments with equal workload as time tends to positive infinity, that is,*

$$\lim_{t\to+\infty} |m_i - m_j| = 0, \quad \forall i, j = 1, ..., n + 1$$

Proof Since $\rho(x, y)$ is continuous on L_μ, $m_i(t)$ is a continuous function with respect to time t. Define

$$\Omega = \{(\lambda_1, \lambda_2, ..., \lambda_n) \in \Delta^n | m_i(t) > 0, i \in E_{n+1}\},$$

where $\Delta = [0, l_b]$. Let k stand for the agent with the least workload such that

$$m_k = \min_{i \in E_{n+1}} \int_{s_{i-1,i}} \rho(x, y)ds.$$

With the control law (2.4), we have

$$\dot{\lambda}_k = m_k - m_{k+1} \le 0$$

$$\dot{\lambda}_{k-1} = m_{k-1} - m_k \ge 0.$$

Therefore, m_k does not decrease as time increases, i.e., $m_k(t) \ge m_k(0) > 0, t \ge 0$. Thus, $m_i(t) \ge m_k(t) \ge m_k(0) > 0, \forall i \in E_{n+1}, t \ge 0$, which implies that Ω is positively invariant with respect to control law (6.2). Take

$$H(x) = \sum_{i=1}^{n+1} m_i^2,$$

and then

$$\dot{H} = -2\sum_{i=1}^{n}(m_i - m_{i+1})^2 \rho(p_{ix}, p_{iy}) \leq 0.$$

Thus, the whole system will converge to the largest invariant set

$$E = \{(\lambda_1, \lambda_2, ..., \lambda_n) \in \Omega | \dot{H} = 0\}$$

with the aid of LaSalle invariance principle. Moreover, since $\rho(x, y) > 0$ on D, $\dot{H} = 0$ is equivalent to $m_i - m_{i+1} = 0, \forall i \in E_n$. Thus, we have

$$m_1 = m_2 = \cdots = m_n = m_{n+1},$$

which completes the proof of this lemma. □

Remark 2.1 Note that $m_i(t) = \int_{s_{i-1,i}} \rho(x, y)ds$ represents the workload on the subsegment $s_{i-1,i}$, $i \in E_{n+1}$. The equilibrium points of n agents are exactly the dividing points of $n + 1$ subsegments with equal workload.

Remark 2.2 Since each agent occupies a certain volume in practice, their initial positions on the segment L_μ cannot be identical. In this sense, the above algorithm is globally convergent.

Suppose the equilibrium point of agent $i \in E_n$ on L_μ is given by $h_i(\mu) = (h_{ix}(\mu), h_{iy}(\mu))^T$. Let $z_i = p_i - h_i(\mu) = (z_{ix}, z_{iy})^T$. Then construct a Lyapunov function

$$V(p) = \sum_{i=1}^{n+1} m_i^2 - \frac{1}{n+1}m^2,$$

where $m = \int_{L_\mu} \rho(x, y)ds$ and $p = (p_1, p_2, ..., p_n) \in R^{2\times n}$. After changing variables $p_i = z_i + h_i(\mu)$, we have $V(p) = V(z, \mu)$, where $z = (z_1, z_2, ..., z_n) \in R^{2\times n}$. Let $z_x = (z_{1x}, z_{2x}, ..., z_{nx})$ and $z_y = (z_{1y}, z_{2y}, ..., z_{ny})$. Furthermore, we obtain the following estimation for $V(z, \mu)$.

Lemma 2.2 *For n agents on L_μ with distinct positions,*

$$\frac{K^2}{(n-1)^2}\|z\|^2 \leq V(z, \mu) \leq 4M^2\|z\|^2 \tag{2.5}$$

holds, when $n \geq 2$. If $n = 1$, then

$$2K^2\|z\|^2 \leq V(z, \mu) \leq 2M^2\|z\|^2 \tag{2.6}$$

where

$$\|z\|^2 = \sum_{i=1}^{n} \|z_i\|^2 = \sum_{i=1}^{n}(\|z_{ix}\|^2 + \|z_{iy}\|^2) = \|z_x\|^2 + \|z_y\|^2.$$

K and M are the minimum and maximum of $\rho(x, y)$ on D, respectively.

Proof Since $h_i(\mu)$ is the equilibrium point of agent i with $i \in E_n$, the following equalities

$$\int_{\mu_x+l_b\cos\theta}^{h_{1x}(\mu)} \hat{\rho}(x)dx = \int_{h_{1x}(\mu)}^{h_{2x}(\mu)} \hat{\rho}(x)dx = \cdots = \int_{h_{nx}(\mu)}^{\mu_x} \hat{\rho}(x)dx$$

hold, where $\hat{\rho}(x) = \rho(x, f(x))$ and $y = f(x)$ stands for the straight-line equation of L_μ. In the case of $n \geq 2$, we simplify $V(z, \mu)$ as follows.

$$\cos^2\theta V(z, \mu) = \left(\int_{h_{1x}(\mu)}^{z_{1x}+h_{1x}(\mu)} \hat{\rho}(x)dx\right)^2 + \sum_{i=2}^{n}\left(\int_{h_{ix}(\mu)}^{z_{ix}+h_{ix}(\mu)} \hat{\rho}(x)dx\right.$$
$$\left. - \int_{h_{(i-1)x}(\mu)}^{z_{(i-1)x}+h_{(i-1)x}(\mu)} \hat{\rho}(x)dx\right)^2 + \left(\int_{h_{nx}(\mu)}^{z_{nx}+h_{nx}(\mu)} \hat{\rho}(x)dx\right)^2.$$

Note that $\sum_{i=1}^{n} x_i^2 \geq \frac{1}{n}(\sum_{i=1}^{n} x_i)^2$, $\forall x_i \in R$. Therefore,

$$\cos^2\theta V(z, \mu) \geq \frac{1}{k}\left(\int_{h_{kx}(\mu)}^{z_{kx}+h_{kx}(\mu)} \hat{\rho}(x)dx\right)^2 \geq \frac{K^2 z_{kx}^2}{k},$$

for $2 \leq k \leq n-1$, and also

$$\cos^2\theta V(z, \mu) \geq \left(\int_{h_{1x}(\mu)}^{z_{1x}+h_{1x}(\mu)} \hat{\rho}(x)dx\right)^2 + \left(\int_{h_{nx}(\mu)}^{z_{nx}+h_{nx}(\mu)} \hat{\rho}(x)dx\right)^2$$
$$\geq K^2(z_{1x}^2 + z_{nx}^2),$$

Accumulating the above terms yields,

$$(n-1)\cos^2\theta V(z, \mu) \geq K^2\left(\sum_{i=1}^{n-1}\frac{1}{i}z_{ix}^2 + z_{nx}^2\right) \geq \frac{K^2}{n-1}\|z_x\|^2.$$

Therefore,

$$\cos^2\theta V(z, \mu) \geq \frac{K^2}{(n-1)^2}\|z_x\|^2,$$

Furthermore, one also has

$$\sum_{i=2}^{n}\left(\int_{h_{ix}(\mu)}^{z_{ix}+h_{ix}(\mu)}\hat{\rho}(x)dx-\int_{h_{(i-1)x}(\mu)}^{z_{(i-1)x}+h_{(i-1)x}(\mu)}\hat{\rho}(x)dx\right)^2$$
$$\leq 2\sum_{i=2}^{n}\left[\left(\int_{h_{ix}(\mu)}^{z_{ix}+h_{ix}(\mu)}\hat{\rho}(x)dx\right)^2+\left(\int_{h_{(i-1)x}(\mu)}^{z_{(i-1)x}+h_{(i-1)x}(\mu)}\hat{\rho}(x)dx\right)^2\right]$$
$$\leq 2M^2\sum_{i=2}^{n}(z_{(i-1)x}^2+z_{ix}^2)$$

Thus,

$$\cos^2\theta V(z,\mu)\leq M^2\left[z_{1x}^2+2\sum_{i=2}^{n}(z_{(i-1)x}^2+z_{ix}^2)+z_{nx}^2\right]\leq 4M^2\|z_x\|^2.$$

We obtain

$$\frac{K^2}{(n-1)^2}\|z_x\|^2\leq\cos^2\theta V(z,\mu)\leq 4M^2\|z_x\|^2. \tag{2.7}$$

For $n=1$,

$$\cos^2\theta V(z,\mu)=2\left(\int_{h_{1x}(\mu)}^{h_{1x}(\mu)+z_{1x}}\hat{\rho}(x)dx\right)^2.$$

Obviously, we have

$$2K^2\|z_x\|^2\leq\cos^2\theta V(z,\mu)\leq 2M^2\|z_x\|^2 \tag{2.8}$$

Since $z_{iy}=\tan\theta z_{ix}, \forall i\in E_n$, we get

$$\frac{K^2}{(n-1)^2}\|z_y\|^2\leq\sin^2\theta V(z,\mu)\leq 4M^2\|z_y\|^2 \tag{2.9}$$

and

$$2K^2\|z_y\|^2\leq\sin^2\theta V(z,\mu)\leq 2M^2\|z_y\|^2 \tag{2.10}$$

hold for $n=1$ and $n\geq 2$, respectively. The inequality (2.7) plus (2.9) equals (2.5), and then we get the inequality (2.6) by (2.8) plus (2.10). The proof is thus completed. □

The preceding lemma provides a bridge between Lyapunov function and partition error, which is useful for error estimation. With regard to the Lyapunov function $V(p)$, we also have the following result.

Lemma 2.3 *For n agents on L_μ with distinct positions, there exists the constant $\gamma_n>0$, such that the inequality*

$$\sum_{i=1}^{n}(m_i - m_{i+1})^2 \geq \gamma_n V(p) \tag{2.11}$$

holds identically, where γ_n is related to n.

Proof For simplicity, we define

$$W = \sum_{i=1}^{n}(m_i - m_{i+1})^2.$$

Note that $m = \sum_{i=1}^{n+1} m_i$ and $m_{n+1} = m - \sum_{i=1}^{n} m_i$. Then we obtain

$$\begin{aligned} W &= \sum_{i=1}^{n}(m_i - m_{i+1})^2 \\ &= \sum_{i=1}^{n-1}(m_i - m_{i+1})^2 + \left(\sum_{i=1}^{n} m_i + m_n - m\right)^2. \end{aligned}$$

Similarly,

$$\begin{aligned} V(p) &= \sum_{i=1}^{n+1} m_i^2 - \frac{m^2}{n+1} \\ &= \sum_{i=1}^{n+1}\left(m_i - \frac{m}{n+1}\right)^2 \\ &= \sum_{i=1}^{n}\left(m_i - \frac{m}{n+1}\right)^2 + \left(\frac{n}{n+1}m - \sum_{i=1}^{n} m_i\right)^2. \end{aligned}$$

By changing variables $x_i = m_i - \frac{m}{n+1}$, $i \in E_n$, it is easy to see that

$$W = \sum_{i=1}^{n-1}(x_i - x_{i+1})^2 + \left(\sum_{i=1}^{n} x_i + x_n\right)^2$$

and

$$V(p) = \sum_{i=1}^{n} x_i^2 + \left(\sum_{i=1}^{n} x_i\right)^2.$$

Obviously, both W and $V(p)$ are positive definite in the quadratic forms with respect to $x = (x_1, x_2, ..., x_n)^T$. Therefore, it is easy to find that there is a constant $\delta > 0$ such that $W \geq \delta \|x\|^2$ holds, and moreover,

$$V(p) \leq (n+1)\|x\|^2$$

due to

$$\|x\|^2 = \sum_{i=1}^{n} x_i^2 \geq \frac{1}{n}\left(\sum_{i=1}^{n} x_i\right)^2, \forall x_i \in R.$$

Thus, $W \geq \frac{\delta}{n+1} V(p)$, which implies the conclusion. □

Remark 2.3 For a given n, γ_n can be obtained by numerical method. In general, γ_n decreases with the increase of n.

In what follows, the real-time workload partition algorithm of region D is discussed and the partition stability of the multi-agent system is also analyzed.

Theorem 2.1 *Assume n agents with distinct initial positions are deployed on the short edge L_b of the rectangular D, on which the partial derivatives of $\rho(x, y)$ with respect to x and y are continuous. If the system of n agents with the following dynamics*

$$\dot{p}_i = \epsilon(\theta)v + \sigma(\theta)u_i, \forall i \in E_n \tag{2.12}$$

where $\epsilon(\theta) = (\sin\theta, -\cos\theta)^T$, moves at common bounded speed $v > 0$ vertical to L_b (see Fig. 2.1) and adopts partition control law (2.4), then the system of n agents is input-to-state stability.

Proof If $v \equiv 0$, the dynamic equation (2.12) reduces to $\dot{p}_i = \sigma(\theta)u_i$. By virtue of Lemma 2.1, n agents on the short edge L_b converge to equilibrium points $h(\mu) = (h_1(\mu), h_2(\mu), ..., h_n(\mu))$ as time tends to positive infinity. We regard $z_i = p_i - h_i(\mu)$ as the state of agent i, and v is viewed as the input of the system. Next, we will consider the relationship between the input v and the state z of the system. The derivative of $V(p)$ with respect to (2.12) is given by,

$$\begin{aligned}
\dot{V}(p) &= \frac{\partial V}{\partial \lambda}\dot{\lambda} + \frac{\partial V}{\partial \mu}\dot{\mu} = \sum_{i=1}^{n}\frac{\partial V}{\partial \lambda_i}\dot{\lambda}_i + \frac{\partial V}{\partial \mu}\dot{\mu} \\
&= \sum_{i=1}^{n}\frac{\partial V}{\partial \lambda_i}\dot{\lambda}_i + \left(\sin\theta\frac{\partial V}{\partial \mu_x} - \cos\theta\frac{\partial V}{\partial \mu_y}\right)v \\
&= -2\sum_{i=1}^{n}(m_{i+1} - m_i)^2\rho(p_{ix}, p_{iy}) + \left(\sin\theta\frac{\partial V}{\partial \mu_x} - \cos\theta\frac{\partial V}{\partial \mu_y}\right)v
\end{aligned} \tag{2.13}$$

where $\lambda = (\lambda_1, \lambda_2, ..., \lambda_n)^T$. Let

$$\Phi = \left\{\frac{k\pi}{2}|k \in \mathbb{Z}\right\}$$

and

$$\rho(p_i) = \rho(p_{ix}, p_{iy}), \quad i = 0, 1, ..., n+1.$$

If $\theta \notin \Phi$, we have

$$\frac{\partial V}{\partial \mu_x} = -2\sec\theta \left[\sum_{i=1}^{n+1} m_i(\rho(p_i) - \rho(p_{i-1})) - \frac{m}{n+1}(\rho(p_{n+1}) - \rho(p_0)) \right]$$
$$+ 2\tan\theta\sec\theta \left[\sum_{i=1}^{n+1} m_i \int_{p_{(i-1)x}}^{p_{ix}} \rho_y(x, f(x))dx - \frac{m}{n+1} \int_{p_{0x}}^{p_{(n+1)x}} \rho_y(x, f(x))dx \right]$$

and

$$\frac{\partial V}{\partial \mu_y} = -2\csc\theta \left[\sum_{i=1}^{n+1} m_i(\rho(p_i) - \rho(p_{i-1})) - \frac{m}{n+1}(\rho(p_{n+1}) - \rho(p_0)) \right]$$
$$+ 2\cot\theta\csc\theta \left[\sum_{i=1}^{n+1} (m_i \int_{p_{(i-1)y}}^{p_{iy}} \rho_x(f^{-1}(y), y)dy) \right.$$
$$\left. - \frac{m}{n+1} \int_{p_{0y}}^{p_{(n+1)y}} \rho_x(f^{-1}(y), y)dy \right].$$

Due to the continuity of $\rho_x(x, y)$ and $\rho_y(x, y)$, both $\frac{\partial V}{\partial \mu_x}$ and $\frac{\partial V}{\partial \mu_y}$ are bounded in D. Hence, there exists a constant $\eta > 0$ satisfying

$$\sin\theta \frac{\partial V}{\partial \mu_x} - \cos\theta \frac{\partial V}{\partial \mu_y} \le \eta.$$

Combining this with Lemma 2.3, we have

$$\dot{V} \le -2K\gamma_n V + \eta v$$

By Comparison Lemma [16], it follows that

$$\begin{aligned} & V(z(t), \mu) \\ & \le V(z(t_0), \mu)e^{-2K\gamma_n(t-t_0)} + \eta \int_{t_0}^{t} e^{2K\gamma_n(\tau - t)} v(\tau)d\tau \\ & \le V(z(t_0), \mu)e^{-2K\gamma_n(t-t_0)} + \frac{\eta}{2K\gamma_n} ||v||_{[t_0,t]} \end{aligned}$$

Based on Lemma 2.2, we obtain

$$\frac{K^2}{(n-1)^2}\|z(t)\|^2 \le 4M^2\|z(t_0)\|^2 e^{-2K\gamma_n(t-t_0)} + \frac{\eta}{2K\gamma_n}||v||_{[t_0,t]}$$

Thus,

$$\|z(t)\| \leq \frac{2(n-1)M}{K} e^{-K\gamma_n(t-t_0)} \|z(t_0)\| + \frac{n-1}{K} \sqrt{\frac{\eta}{2K\gamma_n}} (||v||_{[t_0,t]})^{\frac{1}{2}}$$

for $n \geq 2$, and

$$\|z(t)\| \leq \frac{M}{K} e^{-K\gamma_n(t-t_0)} \|z(t_0)\| + \frac{1}{2K} \sqrt{\frac{\eta}{K\gamma_n}} (||v||_{[t_0,t]})^{\frac{1}{2}}$$

for $n = 1$. Thus, if $\theta \notin \Phi$, we obtain the desired result according to Definition 2.1. Actually, the conclusion still holds when $\theta \in \Phi$ after some manipulations and derivations, which are quite straightforward and omitted here □

Remark 2.4 The condition on the continuity of partial derivatives of $\rho(x, y)$ can be relaxed for $\theta \in \Phi$. Specifically, it is enough to guarantee the continuity of $\rho_y(x, y)$ when $\theta = k\pi, k \in \mathbb{Z}$. In contrast, the continuity of $\rho_y(x, y)$ is needed, for

$$\theta = \frac{2k+1}{2}\pi, \quad k \in \mathbb{Z}.$$

Note that the ISS property is obtained by treating v as the input. In fact, if v is zero (which corresponding to the case when the workload distribution on the moving segment is unchanged when the agents move), then our algorithm guarantees the convergence of equal workload. When v is not zero, the changes of workload distribution keep coming as a disturbance. In other works, v is a measurement of the change rate of the distribution. The larger v, the larger the distribution disturbance will appear in the partition procedure.

2.4 Simulation Results

Before the end of this chapter, numerical simulations are carried out in order to show the effectiveness of the proposed algorithm. For simplicity, system parameters are given as follows: $\theta = \pi$, $n = 3$, $l_a = 10$ and $l_b = 4$. The workload distribution on the rectangular region is denoted by $\rho(x, y) = x + y$, and the initial positions of three agents are (1.5, 0), (3, 0), and (4, 0), respectively.

In Fig. 2.2, blue lines represent borderlines of subregions with accurate workload partition, and red lines stand for the partition trajectories of mobile agents with dynamics (2.12) and sampling period $T_s = 0.01$. Evidently, three agents track the borderlines of accurate workload partition easily after some running time in the simple cases. Additionally, relatively high velocity ($v = 5$) of mobile agents results in larger partition errors on the corresponding segment, which is consistent with theoretical analysis.

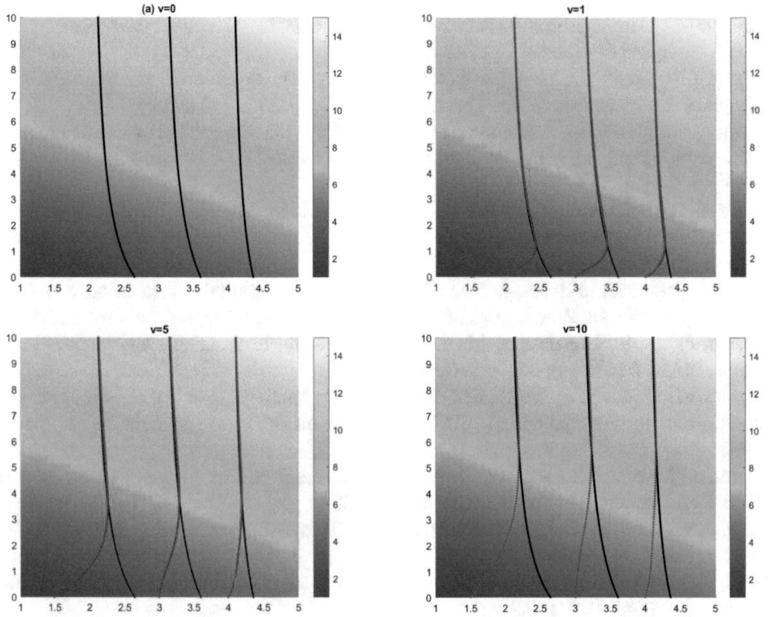

Fig. 2.2 (Color online) Comparison between borderlines of accurate workload partition and trajectories of mobile agents with dynamics (2.12) and various velocities

2.5 Conclusions

This chapter provided a novel online algorithm for partitioning the workload of a given region in a distributed manner. Theoretical analysis was conducted to ensure the ISS of MAS driven by the proposed control algorithm. Numerical simulations are carried out to demonstrate the effectiveness of the online partition approach. Future work may include the consideration of how to employ this online partition approach for efficient sweep coverage of MAS.

References

1. Hong, Y., Hu, J., Gao, L.: Tracking control for multi-agent consensus with an active leader and variable topology. Automatica **42**(7), 1177–1182 (2006)
2. Wang, X., Hong, Y., Jiang, Z.: Coverage tracking of a moving target by a group of mobile agents. In: Proceedings of Asian Control Conference, pp. 27–29 (2009)
3. Olfati-Saber, R., Murray, R.M.: Consensus problems in networks of agents with switching topology and time-delays. IEEE Trans. Autom. Control **49**(9), 1520–1533 (2004)
4. Hussein, I.I., Stipanović, D.M.: Effective coverage control for mobile sensor networks with guaranteed collision avoidance. IEEE Trans. Control Syst. Technol. **15**(4), 642–657 (2007)

5. Gage, D.W.: Command control for many-robot systems. In: AUVS-92, the Nineteenth Annual AUVS Technical Symposium, pp. 22–24 (1992)
6. Cheng, T.M., Savkin, A.V.: A distributed self-deployment algorithm for the coverage of mobile wireless sensor networks. IEEE Commun. Lett. **13**(11), 877–879 (2009)
7. Howard, A., Parker, L., Sukhatme, G.: Experiments with a large heterogeneous mobile robot team: exploration, mapping, deployment and detection. Int. J. Robot. Res. **25**(5), 431–447 (2006)
8. Casbeer, D., Kingston, D., Beard, R., Mclain, T., Li, S., Mehra, R.: Cooperative forest fire surveillance using a team of small unmanned air vehicles. Int. J. Syst. Sci. **37**(6), 351–360 (2006)
9. Schwager, M., McLurkin, J., Rus, D.: Distributed coverage control with sensory feedback for networked robots. In: Proceedings of Robotics: Science and Systems, pp. 49–56 (2006)
10. Du, Q., Faber, V., Gunzburger, M.: Centroidal Voronoi tesseuations: applications and algorithms. SIAM Rev. **41**(4), 637–676 (1999)
11. Pavone, M., Frazzoli, E., Bullo, F.: Distributed policies for equitable partitioning: theory and applications. In: Proceedings of IEEE Conference on Decision and Control, pp. 4191–4197 (2008)
12. Cortés, J., Martínez, S., Karatas, T., Bullo, F.: Coverage control for mobile sensing network. IEEE Trans. Autom. Control **20**(2), 243–255 (2004)
13. Gray, R.M., Neuhoff, D.L.: Quantization. IEEE Trans. Inf. Theory **44**(6), 2325–2383 (1998)
14. Luna, J., Fierro, R., Abdallah, C., Wood, J.: An adaptive coverage control algorithm for deployment of nonholonomic mobile sensors. In: Proceedings of IEEE Conference on Decision and Control, pp. 1250–1256 (2010)
15. Sontag, E.: On the input-to-state stability property. Eur. J. Control. **1**, 24–36 (1995)
16. Khalil, H.K.: Nonlinear Systems, 3rd edn. Prentice Hall (2002)

Chapter 3
Decentralized Cooperative Sweep Coverage Algorithm in Uncertain Environments

3.1 Introduction

The rapid development in information technology makes it possible to achieve desired collective tasks for a large number of interconnected agents, and the research on distributed/decentralized design makes considerable progress and provides feasible ways in various coordination problems [1–4].

In recent years, much attention has been paid to cooperative coverage of multiple agents for its importance in the field of sensor networks, robotic systems, and even social systems. In fact, multi-agent coverage is to assign a group of agents to effectively cover a region of interest or some targets within it in static or dynamic ways. Practical coverage tasks such as search and rescue, exploration, monitoring, and environmental surveillance have been widely investigated with different formulations [5–9]. For example, Voronoi partition was adopted in [5], where the locational cost function was introduced as an index to optimize sensor locations, while the sensor deployment algorithm was designed to monitor random events with a frequency density function and maximize the joint detection probabilities of random events in [6]. Furthermore, an excellent dynamic coverage strategy was developed such that each point in the given domain is effectively covered (with collision avoidance) in [7]. Additionally, an awareness-based model was presented to investigate the coverage control problem over large scale domains in [8].

The sweep coverage as one of the important coverage problems is a dynamic coverage problem [10]. To solve this problem, a group of agents with the sensing capability move across all the given region to detect targets of interest or complete the workload (for example, the dust on the floor) in the region. It is a difficult problem since all the agents have to sweep cooperatively in order to complete the task and even minimize the coverage time. Reference [11] proposed a formation-based sweep coverage strategy, while [12] discussed the coverage formation under the requirement of the minimum dwell time. In addition, [13] proposed a decomposition method for the coverage of a known area. Reference [14] presented a distributed sensor-based coverage algorithm for a team of square robots to cover a finite rectangular envi-

C. Zhai et al., *Cooperative Coverage Control of Multi-Agent Systems and its Applications*, Studies in Systems, Decision and Control 408,
https://doi.org/10.1007/978-981-16-7625-3_3

ronment. Reference [15] reported a heuristic algorithm for multiple robots, which cooperatively sweep an area with obstacles with the help of the market-based negotiation mechanism. Inspired by the behavior of insects, [16] designed the ant robots with smell traces to complete the coverage of a given complex environment. In fact, [17] provided a good survey on sweep coverage problems.

The objective of this chapter is to discuss the dynamic sweep coverage with uncertain workload density. Most of the existing approaches were obtained for the uniform/known workload distribution, which is not applicable to the case of uncertain/nonuniform workload distribution. Due to the uncertainties of the nonuniform workload, it is impossible to give a fixed coverage strategy for each agent to achieve the minimum coverage time in advance. Moreover, the scheme of dynamic workload assignment is adopted to learn the uncertainties online. The main contributions of the paper include: propose the decentralized sweep coverage algorithm with combined two operations: workload partition (to handle the uncertainties based on sensors equipped in agents) and sweep (to complete the coverage related to actuators in agents); estimate an upper bound of the extra time (that is, the difference between the actual coverage time and the optimal time) resulting from workload uncertainties.

The rest of this chapter is organized as follows. Section 3.2 presents a new sweep coverage formulation in a region with unknown workload and then proposes a decentralized sweep coverage algorithm. Then Sect. 3.3 provides a theoretical estimation of the extra time to complete the coverage, followed by numerical simulations. Finally, Sect. 3.4 gives concluding remarks.

3.2 Formulation and Coverage Algorithm

In this section, we give a formulation for n mobile agents to sweep a given region with unknown workload distribution and then propose a decentralized control algorithm of MAS.

As we know, to increase the coverage effectiveness, the whole region can simply be divided into n connected subregions, and each agent is responsible for the coverage task on its own subregion to complete the sweep coverage in a cooperative way. If the workload density in a covered region is known for all the agents in advance, the optimal strategy to complete the coverage may be carried out by partitioning the whole region into subregions with equal workload for all the agents so as to complete the sweeping task at the same time. Nevertheless, the limited sensing range of each agent and the unknown workload distribution make it impossible to assign average workload to each agent offline before any action. Therefore, we have to provide online coverage algorithms.

We first consider a bounded and closed set D enclosed by two parallel lines (corresponding to $y = 0$ and $y = h$, respectively) and two continuous curves described by $x = g_a(y)$ and $x = g_b(y)$ with $g_b(y) > g_a(y)$ as shown in Fig. 3.1. Later, we will discuss to cover a generalized set S. In our problem, to focus on the workload partition and sweeping operation we assume all the agents line up at the bottom of

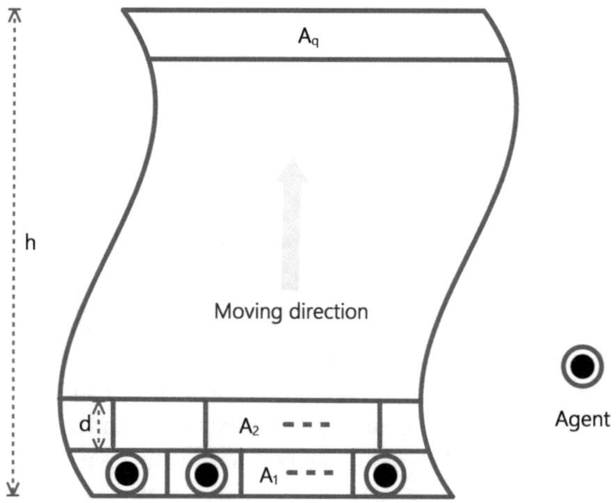

Fig. 3.1 Sweep coverage of multiple mobile agents on the region D

region D (regarded as the start line of the agents) as their initial positions for simplicity, and they will move to the top (as the terminal line) of D by sweeping all the bounded region. Actually, for the agents with random initial positions, we can employ the distributed algorithm to drives all agents to line up first as done in the sweep coverage control [11]. Suppose each agent has its actuation range with the diameter d, and therefore, when it sweeps, it will clean up a stripe with the width d. Usually $d << h$, and approximately the whole region D is partitioned into stripes with the same width d. For convenience, we assume $h = qd$ for some integer $q > 0$. Then the agents sweep these stripes one by one for the coverage of D.

To shorten the whole time for coverage, each stripe is also partitioned into sub-stripes for each agent so that each agent has the same workload. Furthermore, each agent can measure the workload around its position. Then each agent sweeps its own sub-stripe and simultaneously partitions the next stripe according to the proposed workload partition algorithm. If some agent completes the workload on its own sub-stripe, it stops the sweeping operation, but continues the partition operation with the help of neighbor partition information. Once the workload on the current stripe is completed, the partition operation on the next stripe stops and all the agents start to sweep their own newly-partitioned sub-stripes on the next stripe, and meanwhile partition the follow-up stripe. The procedure keeps repeated until D is swept. The workload distribution is denoted by $\rho(x, y)$, which is continuous and uncertain on D. We give the following assumption for $\rho(x, y)$.

Assumption There exist positive constants $\bar{\rho}$ and $\underline{\rho}$ such that

$$\underline{\rho} \leq \rho(x, y) \leq \bar{\rho}, \quad \forall (x, y) \in D.$$

Denote the vertical range of the kth stripe, A_k, as $[y_{k-1}, y_k)$ with $y_0 = 0$, $y_q = h$ and $y_k = y_{k-1} + d$ for $k = 1, ..., q$. Then the workload distribution on the kth stripe becomes

$$\omega_k(\tau) = \int_{y_{k-1}}^{y_k} \rho(\tau, y)dy, \quad 1 \le k \le q.$$

where $\rho(x, y) = 0, \forall(x, y) \notin D$. Thus,

$$m^k = \int_{x_0}^{x_n} \omega_k(\tau)d\tau$$

denotes the workload on stripe A_k with $x_0 = \min_{(x,y)\in D} x$, and $x_n = \max_{(x,y)\in D} x$. Since the positions x_0 and x_n keep unchanged during the partitioning, we have $\dot{x}_0 = \dot{x}_n = 0$. On each stripe, there are n sub-stripes for n agents. Correspondingly, there are $n - 1$ partition marks to separate the n sub-stripes. In other words, partition mark i is the common boundary of sub-stripes i and $i + 1$ (swept by agents i and $i + 1$, respectively). The workload on the ith sub-stripe of A_k is given by

$$m_i^k = \int_{x_{i-1}}^{x_i} \omega_k(\tau)d\tau \tag{3.1}$$

where x_{i-1} and x_i are horizontal positions of partition marks $i - 1$ and i on A_k, respectively, with $i \in E_n = \{1, 2, ..., n\}$.

For simplicity, we assume each agent can only communicate with its nearest neighbors and it has the same sweeping rate ν (i.e., the amount of workload completed by each agent in unit time). Obviously, the completion time of sweeping each sub-stripe only depends on ν and workload on the sub-stripe. Clearly, if $\rho(x, y)$ is known in advance, we can partition the workload and complete the sweeping task in the shortest time. Denote T_{opt} as the shortest/optimal coverage time of D, which can be easily obtained as follows

$$T_{opt} = \frac{1}{n\nu} \iint_D \rho(x, y)dxdy \tag{3.2}$$

However, due to the uncertain workload density, we cannot make the workload partition beforehand, and we have to carry out the partition operation online.

Our algorithm contains two basic operations: partition the workload on each stripe and sweep stripe by stripe. In fact, the partitioning operation on a stripe (say A_k) for ith agent can be implemented as follows:

- Measure the workload on ith and $(i + 1)$th sub-stripes (i.e., m_i^k and m_{i+1}^k);
- Update the ith partition mark, x_i (that is the mark between ith and $(i + 1)$th sub-stripes), as follows:

$$\dot{x}_i = \kappa(m_{i+1}^k - m_i^k), \quad i = 1, ..., n - 1 \tag{3.3}$$

for a given constant $\kappa > 0$.

Clearly, the partition marks become boundaries of sub-stripes when the partition operation on a given stripe stops. In addition, the initial positions of partition marks on stripe A_{k+1} correspond to the final positions of partition marks on A_k. Moreover, the agents will move together from one stripe to the next one for sweeping operation.

Remark 3.1 Distributed protocols can be designed to guarantee that all the agents move from one stripe to the next (say, A_k to A_{k+1}) simultaneously. For example, each agent exchanges the workload information of sub-stripes with its adjacent neighbors to get the the most workload on the sub-stripes

$$m_*^k = \max_{1 \le i \le n} m_i^k.$$

There were distributed algorithms to find the optimal values of agents such as the classic LCR (Le Lann-Chang-Roberts) algorithm [2]. With a modified LCR algorithm, each agent can get m_*^k in the distributed way. Omitting the detailed sweep procedure, we can roughly regard the completion time of sweeping stripe A_k as

$$T_k = \frac{m_*^k}{\nu} = \frac{1}{\nu} \max_{1 \le i \le n} m_i^k \tag{3.4}$$

which can be taken as the common switching time for all the agents to move from A_k to A_{k+1}. In other words, the completion time to sweep A_k is related to the sweeping rate ν and m_*^k. In addition, T_k can be computed by each agent since ν is known. Hence, all the agents can move together to A_{k+1} from A_k based on their own computation of T_k after the coverage on A_k.

Here we propose our coverage algorithm, Decentralized Sweep Coverage Algorithm (DSCA), in Table 3.1. To make our algorithm feasible, we need to assume that the partition marks on each stripe will not intersect with lateral boundaries of the region D during coverage in what follows. In fact, the assumption can be guaranteed if the workload near the lateral boundaries of each stripe can be sufficiently small. In addition, the assumption obviously holds if D is rectangular. Thus, the total coverage time of the region D is $T^* = \sum_{k=1}^{q} T_k$ based on (3.4). The objective of our research is to estimate the coverage time T^* for its potential minimization, or equivalently estimate the extra time between the coverage time and the optimal one, that is

$$\Delta T = T^* - T_{opt}.$$

Since it is hard to get the exact expression of ΔT, we focus on an estimation of ΔT for the proposed sweep coverage algorithm.

3.3 Technical Analysis

In this section, the theoretical analysis for the coverage algorithm DSCA is provided.

Table 3.1 Decentralized sweep coverage algorithm

Goal: sweep the given region

For $i \in E_n$, i-th agent performs as follows.

set $k = 1$
while A_k is not the last stripe on the given region **do**
  while m^k is not completed **do**
    if m_i^k is not completed **then**
      update the i-th partition mark on A_{k+1} and sweep i-th sub-stripe on A_k
    else
      update the i-th partition mark on A_{k+1}
    end if
  end while
  stop partition operation and move to i-th sub-stripe on A_{k+1}
  set $k = k + 1$
end while
sweep i-th sub-stripe on A_q until the workload is finished

3.3.1 Key Lemmas

In this subsection, we give three lemmas before introducing main results. Here is the first result.

Lemma 3.1

$$\lim_{t \to +\infty} |m_i^k - m_j^k| = 0, \forall i, j \in E_n$$

with m_i^k defined in (3.1).

Proof Take

$$\begin{aligned} H_k = \sum_{i=1}^{n} \left(m_i^k - \frac{m^k}{n} \right)^2 &= \sum_{i=1}^{n} [(m_i^k)^2 + \frac{(m^k)^2}{n^2} - \frac{2m^k}{n} \cdot m_i^k] \\ &= \sum_{i=1}^{n} (m_i^k)^2 + \sum_{i=1}^{n} \frac{(m^k)^2}{n^2} - \frac{2m^k}{n} \sum_{i=1}^{n} m_i^k \\ &= \sum_{i=1}^{n} (m_i^k)^2 + n \cdot \frac{(m^k)^2}{n^2} - \frac{2m^k}{n} \cdot m^k \\ &= \sum_{i=1}^{n} (m_i^k)^2 - \frac{1}{n} (m^k)^2 \end{aligned} \tag{3.5}$$

as a Lyapunov function candidate on the stripe A_k to describe how uniform the workload partition on A_k is. Since

$$m^k = \sum_{i=1}^{n} m_i^k$$

is constant, we have

$$\dot{m}^k = \sum_{i=1}^{n} \dot{m}_i^k = 0.$$

Obviously,

$$\dot{m}_i^k = \omega_k(x_i)\dot{x}_i - \omega_k(x_{i-1})\dot{x}_{i-1}, i \in E_n,$$

and particularly,

$$\dot{m}_1^k = \omega_k(x_1)\dot{x}_1 - \omega_k(x_0)\dot{x}_0 = \omega_k(x_1)\dot{x}_1$$

and

$$\dot{m}_n^k = \omega_k(x_n)\dot{x}_n - \omega_k(x_{n-1})\dot{x}_{n-1} = -\omega_k(x_{n-1})\dot{x}_{n-1}.$$

Clearly, the derivative of H_k along the trajectories of the system satisfies

$$\begin{aligned}
\dot{H}_k &= 2\sum_{i=1}^{n} m_i^k \cdot \dot{m}_i^k - \frac{2m^k}{n}\dot{m}^k \\
&= 2\sum_{i=1}^{n} m_i^k \cdot \dot{m}_i^k \\
&= 2\sum_{i=1}^{n} (\omega_k(x_i)\dot{x}_i - \omega_k(x_{i-1})\dot{x}_{i-1})m_i^k \\
&= 2\sum_{i=1}^{n-1} (m_i^k - m_{i+1}^k)\omega_k(x_i)\dot{x}_i \\
&= -2\kappa\sum_{i=1}^{n-1} (m_{i+1}^k - m_i^k)^2\omega_k(x_i)
\end{aligned}$$

Since $\kappa > 0$ and

$$\omega_k(x_i) \geq d\underline{\rho} > 0,$$

$\dot{H}_k < 0$ on the stripe A_k except for the equilibrium point, which corresponds to

$$m_1^k = m_2^k = \cdots = m_n^k.$$

Thus the equilibrium point is asymptotically stable, which implies the conclusion. □

Lemma 3.1 shows that, the partitioning algorithm given in (3.3) can yield the optimal partition if it runs for infinite time. However, it only runs for a finite time. The length of time the partitioning algorithm runs is actually the amount of time spent by the agent with the most workload. Next, we have

Lemma 3.2 *With H_k defined in (3.5) on stripe A_k,*

$$\frac{\lambda_{min}}{n} H_k \leq \sum_{i=1}^{n-1} (m_{i+1}^k - m_i^k)^2 \leq \lambda_{max} H_k$$

where λ_{min} and λ_{max} are the minimum and maximum eigenvalues of the positive definite matrix Γ as follows:

$$\Gamma = \begin{pmatrix} 2\,0\,1\,.\,.\,1\,1\,2 \\ 0\,3\,0\,.\,.\,1\,1\,2 \\ 1\,0\,3\,.\,.\,1\,1\,2 \\ \cdot\;\cdot\;\cdot\;\cdot\;\cdot\;\cdot\;\cdot\;\cdot \\ \cdot\;\cdot\;\cdot\;\cdot\;\cdot\;\cdot\;\cdot\;\cdot \\ 1\,1\,1\,.\,.\,3\,0\,2 \\ 1\,1\,1\,.\,.\,0\,3\,1 \\ 2\,2\,2\,.\,.\,2\,1\,5 \end{pmatrix} \in R^{(n-1)\times(n-1)},$$

for $n > 2$ or $\Gamma = 4$ for $n = 2$.

Proof Take $\delta_i = m_i^k - m^k/n, i \in E_n$, which yields

$$\sum_{i=1}^{n} \delta_i = 0$$

or

$$\delta_n = -\sum_{i=1}^{n-1} \delta_i.$$

Then we can get

$$
\begin{aligned}
\sum_{i=1}^{n-1}(m_{i+1}^k - m_i^k)^2 &= \sum_{i=1}^{n-1}(\delta_{i+1} - \delta_i)^2 \\
&= \sum_{i=1}^{n-2}(\delta_{i+1} - \delta_i)^2 + \left(\sum_{i=1}^{n-1}\delta_i + \delta_{n-1}\right)^2 \\
&= \sum_{i=1}^{n-2}(\delta_{i+1} - \delta_i)^2 + \left(\sum_{i=1}^{n-2}\delta_i + 2\delta_{n-1}\right)^2 \\
&= 2\delta_1^2 + 3\sum_{i=2}^{n-2}\delta_i^2 + 5\delta_{n-1}^2 - 2\sum_{i=1}^{n-2}\delta_i\delta_{i+1} \\
&+ 2\sum_{1\le i<j\le n-2}\delta_i\delta_j + 4\delta_{n-1}\sum_{i=1}^{n-2}\delta_i \\
&= 2\delta_1^2 + 3\sum_{i=2}^{n-2}\delta_i^2 + 5\delta_{n-1}^2 + 2\delta_{n-2}\delta_{n-1} \\
&+ 2\sum_{1\le i<j\le n-2, j\neq i+1}\delta_i\delta_j + 4\delta_{n-1}\sum_{i=1}^{n-3}\delta_i \\
&= \delta^T\Gamma\delta
\end{aligned}
$$

and

$$H_k = \|\delta\|^2 + \left(\sum_{i=1}^{n-1}\delta_i\right)^2$$

where $\delta^T = (\delta_1, \delta_2, ..., \delta_{n-1})$ and $\|\cdot\|$ is the Euclidean norm. Since

$$\lambda_{min}\|\delta\|^2 \le \delta^T\Gamma\delta \le \lambda_{max}\|\delta\|^2$$

and

$$\|\delta\|^2 \le H_k \le n\|\delta\|^2,$$

we obtain

$$\frac{\lambda_{min}}{n}H_k \le \delta^T\Gamma\delta \le \lambda_{max}H_k.$$

The proof is thus completed. □

Remark 3.2 λ_{min} and λ_{max} can be computed with the aid of numerical methods provided in [18] in practice.

Denote $\hat{m}_i^k$ and $\tilde{m}_i^k$ as the workload on the ith sub-stripe of A_k with the initial and final positions of partition marks, respectively. Take

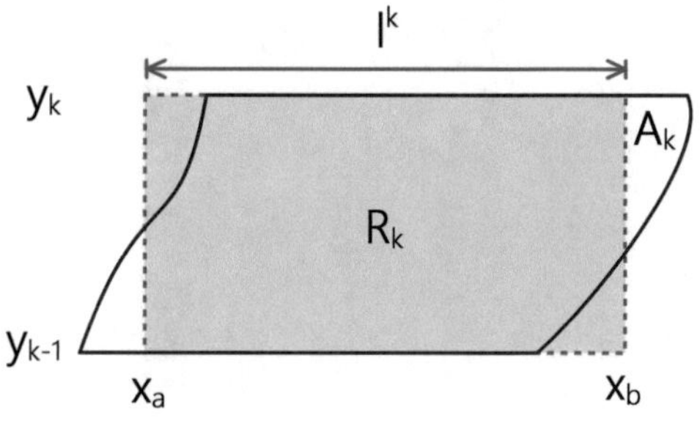

Fig. 3.2 The transformation from the stripe A_k to the rectangular region R_k

$$H_k^a = \sum_{i=1}^{n}\left(\hat{m}_i^k - \frac{m^k}{n}\right)^2 = \sum_{i=1}^{n}(\hat{m}_i^k)^2 - \frac{1}{n}(m^k)^2$$

and

$$H_k^b = \sum_{i=1}^{n}\left(\tilde{m}_i^k - \frac{m^k}{n}\right)^2 = \sum_{i=1}^{n}(\tilde{m}_i^k)^2 - \frac{1}{n}(m^k)^2,$$

which denote the uniformity of the workload partition at the initial and final partition positions on A_k, respectively. Note that the boundaries between sub-stripes on A_k are exactly the final positions of the partition marks on the stripe. Clearly, the initial positions of partition marks on A_{k+1} are the final partition positions on A_k. Therefore, we obtain the next lemma.

Lemma 3.3 *H_{k+1}^a and H_k^b satisfy the following inequality*

$$H_{k+1}^a \le \alpha_d^2 H_k^b + \beta_d \frac{(dl^k)^2}{n}, \quad k = 1, ..., q-1 \tag{3.6}$$

with

$$l^k = \frac{1}{d}\int_{y_{k-1}}^{y_k} (g_b(y) - g_a(y))dy,$$

and

$$\alpha_d = \tau_d \frac{\bar{\rho}}{\underline{\rho}}, \quad \beta_d = \tau_d^2 \frac{\bar{\rho}^4}{\underline{\rho}^2} - \sigma_d^2 \underline{\rho}^2, \quad \sigma_d = \min_{1 \le k \le q} \frac{l^{k+1}}{l^k},$$

where τ_d is a constant related to d, $g_a(y)$ and $g_b(y)$.

Proof For convenience, we consider to construct the equivalent rectangular region R_k from the stripe A_k with the guarantee that both the area and the workload on the corresponding sub-stripes remain unchanged. Here, R_k is enclosed by four lines, corresponding to $y = y_{k-1}$, $y = y_k$, $x = x_a$ and $x = x_b$ (referring to Fig. 3.2), where

$$x_a = \frac{1}{d}\int_{y_{k-1}}^{y_k} g_a(y)dy, \quad x_b = \frac{1}{d}\int_{y_{k-1}}^{y_k} g_b(y)dy.$$

Clearly, the workload distribution on the region $A_k \bigcap R_k$ is $\rho(x, y)$. Define

$$\Omega_a = \{(x, y)|(x, y) \in A_k, (x, y) \notin R_k, x \le x_a\}$$

and

$$\bar{\Omega}_a = \{(x, y)|(x, y) \notin A_k, (x, y) \in R_k, x_a \le x \le g_a(y)\}.$$

Similarly, let

$$\Omega_b = \{(x, y)|(x, y) \in A_k, (x, y) \notin R_k, x \ge x_b\}$$

and

$$\bar{\Omega}_b = \{(x, y)|(x, y) \notin A_k, (x, y) \in R_k, g_b(y) \le x \le x_b\}$$

Define the workload distribution on the region $\bar{\Omega}_a \bigcup \bar{\Omega}_b$ as $\mu(x, y) \in [\underline{\rho}, \bar{\rho}]$ with the constraints

$$\iint_{\Omega_a} \rho(x, y)dxdy = \iint_{\bar{\Omega}_a} \mu(x, y)dxdy, \quad \iint_{\Omega_b} \rho(x, y)dxdy = \iint_{\bar{\Omega}_b} \mu(x, y)dxdy.$$

Since any partition marks do not intersect with lateral boundaries of D during coverage, both the area and the workload on the corresponding sub-stripes keep unchanged after the above transformation. Hence, H_k is also invariant to the transformation. Clearly, $l^k = x_b - x_a$ is the length of kth rectangular stripe R_k with workload m^k, and let $\tilde{l}_i^k$ $(1 \le i \le n)$ be the length of ith sub-stripe on R_k, whose boundaries correspond to the final positions of $i-1$th and ith partition marks. Then $l^{k+1} \ge \sigma_d l^k$, $\forall 1 \le k \le q$ with

$$\sigma_d = \min_{1 \le k \le q} \frac{l^{k+1}}{l^k}.$$

Considering the workload on each sub-stripe of R_k, we have

$$(\underline{\rho}\tilde{l}_i^k)^2 \le \frac{1}{d^2}(\tilde{m}_i^k)^2 = \frac{1}{d^2}\left(\int_{x_{i-1}}^{x_i} \omega_k(\tau)d\tau\right)^2 \le (\bar{\rho}\tilde{l}_i^k)^2$$

Since the positions of partition marks on A_k are always the same as those on R_k, the initial partition positions on R_{k+1} also correspond to the boundaries between subregions on R_k. Denote $\hat{l}_i^{k+1}$ as the length of the ith sub-stripe formed by the initial positions of $i-1$th and ith partition marks on R_{k+1}. Then $\hat{l}_i^{k+1} = \tilde{l}_i^k$ $(2 \le i \le (n-1))$ and $\hat{l}_i^{k+1} \le \tau_d \tilde{l}_i^k$ with

$$\tau_d = \max_{1 \le k \le q, i \in E_n} \frac{\hat{l}_i^{k+1}}{\tilde{l}_i^k}$$

Thus, we have

$$(\hat{m}_i^{k+1})^2 \leq (d\bar{\rho}\hat{l}_i^{k+1})^2 \leq \tau_d^2 \frac{\bar{\rho}^2}{\underline{\rho}^2} (\tilde{m}_i^k)^2$$

Accumulating the n terms on both R_k and R_{k+1} yields,

$$\sum_{i=1}^{n} (\hat{m}_i^{k+1})^2 \leq \alpha_d^2 \sum_{i=1}^{n} (\tilde{m}_i^k)^2 \tag{3.7}$$

Similarly, $(dl^k \underline{\rho})^2 \leq (m^k)^2 \leq (dl^k \bar{\rho})^2$. Due to $l^{k+1} \geq \sigma_d l^k$, we obtain

$$(m^{k+1})^2 \geq (dl^{k+1}\underline{\rho})^2 \geq \left(\frac{l^{k+1}\underline{\rho}}{l^k \bar{\rho}} m^k \right)^2 \geq \sigma_d^2 \frac{\underline{\rho}^2}{\bar{\rho}^2} (m^k)^2$$

and

$$-\frac{(m^{k+1})^2}{n} \leq -\frac{\sigma_d^2 \underline{\rho}^2}{n \bar{\rho}^2} (m^k)^2 \tag{3.8}$$

Combining (3.7) with (3.8) gives

$$H_{k+1}^a \leq \alpha_d^2 H_k^b + \frac{\tau_d^2 \frac{\bar{\rho}^2}{\underline{\rho}^2} - \sigma_d^2 \frac{\underline{\rho}^2}{\bar{\rho}^2}}{n} (m^k)^2 \leq \alpha_d^2 H_k^b + \beta_d \frac{(dl^k)^2}{n}$$

which completes the proof of this lemma. □

Remark 3.3 Generally, $\tau_d \geq 1$ because $\hat{l}_i^{k+1} = \tilde{l}_i^k, 2 \leq i \leq (n-1)$, and τ_d tends to 1 as the width of stripes d goes to 0. Moreover, we can prove $\beta_d \geq 0$. Clearly, $\tau_d = \sigma_d = 1$ if the region D is a rectangle.

3.3.2 Main Results

In this subsection, we discuss two cases: the region D and a generalized region S. The final goal is to complete the coverage task as soon as possible. If all the environmental information is known in advance, we can give the optimal partition offline, and then achieve the optimal coverage time T_{opt} defined in (3.2). However, because of the workload uncertainties, we have to conduct the partition online during the sweep motion, and the unequal workload partitioned in finite time renders more time to complete the whole sweep coverage. Thus, we estimate the extra time to cover D in addition to the optimal coverage time. The following is our main result.

Theorem 3.1 *With the DSCA given in Table 3.1, the extra time to sweep the region D spent in addition to the optimal coverage time is bounded by*

$$\Delta T \leq \frac{\sqrt{H_1^b}}{\nu}\left(1+\sum_{k=1}^{q-1}\alpha_d^k e^{-\frac{c}{n}\sum_{i=1}^{k}t_i}\right)+\frac{d}{\nu}\sqrt{\frac{\beta_d}{n}}\sum_{k=1}^{q-1}l^k\left(\sum_{i=0}^{q-1-k}\alpha_d^i e^{-\frac{c}{n}\sum_{j=k}^{k+i}t_j}\right) \tag{3.9}$$

with

$$c = \lambda_{min}\underline{\rho}\kappa d$$

and

$$t_k = \frac{\underline{\rho}}{n\nu}\int_{y_{k-1}}^{y_k}(g_b(y)-g_a(y))dy.$$

Proof Since the agents sweep their own sub-stripes simultaneously, the total time spent on sweeping the whole stripe is determined by the heaviest sub-stripe workload m_*^k. Obviously, the optimal partition of the stripe yields sub-stripes with equal workload. We first estimate the difference between the most workload and average workload on the subregion of each stripe. Then the extra time to sweep each stripe compared to the optimal partition is calculated. Finally, we obtain an upper bound of the extra time of sweeping D. The time derivative of H_k with respect to (3.3) is given by

$$\dot{H}_k = -2\kappa\sum_{i=1}^{n-1}(m_{i+1}^k - m_i^k)^2\omega_k(x_i)$$

From Lemma 3.2, we have

$$\sum_{i=1}^{n-1}(m_{i+1}^k - m_i^k)^2\omega_k(x_i) \geq \frac{\lambda_{min}}{n}\omega_k(x_i)H_k \geq \frac{\lambda_{min}d\underline{\rho}}{n}H_k$$

Thus,

$$\dot{H}_k \leq -2\frac{\lambda_{min}\kappa d\underline{\rho}}{n}H_k \tag{3.10}$$

Solving (3.10) gives

$$H_k(t) \leq H_k^a e^{-2\frac{\lambda_{min}\kappa d\underline{\rho}}{n}t}$$

The time spent on partitioning the stripe A_{k+1} is exactly the time to sweep the stripe A_k, which has the lower bound

$$t_k = \frac{d\underline{\rho}l^k}{n\nu} = \frac{\underline{\rho}}{n\nu}\int_{y_{k-1}}^{y_k}(g_b(y)-g_a(y))dy.$$

Therefore,

$$H_{k+1}^b \leq H_{k+1}(t_k) \leq H_{k+1}^a e^{-2\frac{\lambda_{min}\kappa d\underline{\rho}}{n}t_k}$$

Substituting (3.6) in Lemma 3.3 into the above inequality, we get

$$H_{k+1}^b \leq \left(\alpha_d^2 H_k^b + \beta_d \frac{(dl^k)^2}{n}\right) e^{-2\frac{\lambda_{min}\kappa d \underline{\rho}}{n} t_k} \tag{3.11}$$

As a result,

$$\sqrt{H_{k+1}^b} \leq \left(\alpha_d \sqrt{H_k^b} + dl^k \sqrt{\frac{\beta_d}{n}}\right) e^{-\frac{\lambda_{min}\kappa d \underline{\rho}}{n} t_k}$$

Hence, the extra time to sweep the whole region D is bounded by

$$\begin{aligned}\Delta T &\leq \frac{1}{v}\sum_{k=1}^{q}\sqrt{H_k^b} \\ &\leq \frac{\sqrt{H_1^b}}{v}\left(1+\sum_{k=1}^{q-1}\alpha_d^k e^{-\frac{c}{n}\sum_{i=1}^k t_i}\right) + \frac{d}{v}\sqrt{\frac{\beta_d}{n}}\sum_{k=1}^{q-1} l^k \left(\sum_{i=0}^{q-1-k}\alpha_d^i e^{-\frac{c}{n}\sum_{j=k}^{k+i} t_j}\right)\end{aligned}$$

which completes the proof. □

Based on Theorem 3.1 and (3.2), the coverage time with the proposed algorithm becomes

$$\begin{aligned}T^* \leq& \frac{1}{nv}\iint_D \rho(x,y)dxdy + \frac{\sqrt{H_1^b}}{v}\left(1+\sum_{k=1}^{q-1}\alpha_d^k e^{-\frac{c}{n}\sum_{i=1}^k t_i}\right) \\ &+ \frac{d}{v}\sqrt{\frac{\beta_d}{n}}\sum_{k=1}^{q-1} l^k \left(\sum_{i=0}^{q-1-k}\alpha_d^i e^{-\frac{c}{n}\sum_{j=k}^{k+i} t_j}\right)\end{aligned}$$

Remark 3.4 From the inequality (3.9), more accurate estimation on the extra time can be obtained when we increase the value of the constant κ. The first term of (3.9) is an error caused by the initial partition error H_1^b, while the second term mainly results from the non-uniform workload distribution and the irregular boundaries. If we partition the first stripe into n sub-stripes with equal workload (that is, $H_1^b = 0$), the first term will disappear. Hence, ΔT will tend to 0 as the number of agents, n, goes to infinity.

Remark 3.5 In the practical computation, κ can not be chosen arbitrarily. Assume that Euler method is adopted to implement the partition algorithm. To ensure the estimation accuracy of (3.9), κ should be chosen such that

$$\frac{\kappa d \bar{\rho}^3}{4\underline{\rho} v}\left(\frac{1-\gamma^q}{1-\gamma} - q\right) T_s$$

is sufficiently small. Here,

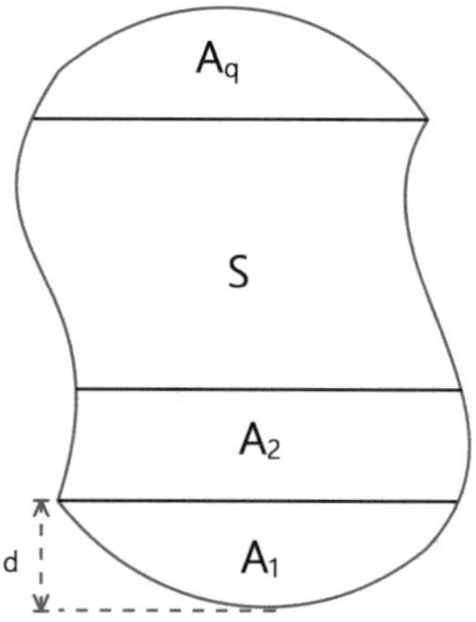

Fig. 3.3 A generalized region S to be covered

$$\gamma = e^{\frac{16\kappa d^2 \bar{\rho}^2 l_{max}}{\nu}}, \quad l_{max} = \max_{1 \le k \le q} l^k$$

and T_s denotes the sampling period or step size. Moreover, it also satisfies the inequality $2\kappa\bar{\rho}dT_s - 1 < 0$ so that the partition marks do not collide with each other. With other numerical methods such as Runge–Kutta method, we can make the constraint on κ less conservative.

Remark 3.6 For the rectangular region D with width l and length h, we have $\tau_d = \sigma_d = 1$ and $l^k = l$. Then (3.9) can be reduced to

$$\Delta T \le \frac{(1 - \bar{\rho}^q e^{-\frac{qcdl\underline{\rho}}{n^2\nu}}/\underline{\rho}^q)\sqrt{H_1^b}}{(1 - \bar{\rho}e^{-\frac{cdl\underline{\rho}}{n^2\nu}}/\underline{\rho})\nu} + \frac{dl}{\nu}\sqrt{\frac{\bar{\rho}^4 - \underline{\rho}^4}{n\underline{\rho}^2}} \sum_{j=1}^{q-1} j e^{-\frac{(q-j)cdl\underline{\rho}}{n^2\nu}} \left(\frac{\bar{\rho}}{\underline{\rho}}\right)^{q-1-j} \quad (3.12)$$

with $c = \lambda_{min}\underline{\rho}\kappa d$.

Actually, our algorithm can also be applied to a generalized bounded region S (see Fig. 3.3). In fact, the given region S is divided into q stripes with the width d, and it is composed of three parts: A_1, $\bigcup_{k=2}^{q-1} A_k$ and A_q, where $\bigcup_{k=2}^{q-1} A_k$ is a region with two parallel edges (like D). Thus, the extra time to sweep each part can be estimated. The following result is given for the region S.

Theorem 3.2 *With the* DSCA, *the extra time to sweep the given region S spent in addition to the optimal coverage time is bounded by*

$$\begin{aligned}\Delta T \le{}& \frac{\bar{\rho}d(n-1)(\bar{l}_1 + \bar{l}_q)}{n\nu} + \frac{\sqrt{H_2^b}}{\nu}\left(1 + \sum_{k=1}^{q-3} \alpha_d^k e^{-\frac{c}{n}\sum_{i=2}^{k+1} t_i}\right) \\ &+ \frac{d}{\nu}\sqrt{\frac{\beta_d}{n}} \sum_{k=2}^{q-2} l^k \left(\sum_{i=0}^{q-2-k} \alpha_d^i e^{-\frac{c}{n}\sum_{j=k}^{k+i} t_j}\right)\end{aligned} \quad (3.13)$$

with

$$\bar{l}_k = \max_{(x,y)\in A_k} x - \min_{(x,y)\in A_k} x$$

for $k = 1$ *or* q.

Proof Denote the extra time to sweep the ith stripe on the region S as Δt_i for $1 \le i \le q$. From Theorem 3.1, the extra time to sweep the region $\bigcup_{k=2}^{q-1} A_k$ with parallel start and terminal lines is bounded by

$$\sum_{k=2}^{q-1} \Delta t_k \le \frac{\sqrt{H_2^b}}{\nu}\left(1 + \sum_{k=1}^{q-3} \alpha_d^k e^{-\frac{c}{n}\sum_{i=2}^{k+1} t_i}\right) + \frac{d}{\nu}\sqrt{\frac{\beta_d}{n}} \sum_{k=2}^{q-2} l^k \left(\sum_{i=0}^{q-2-k} \alpha_d^i e^{-\frac{c}{n}\sum_{j=k}^{k+i} t_j}\right)$$

In addition, the extra time to sweep the stripe A_1 satisfies

$$\Delta t_1 = \frac{1}{\nu}(\max_{1\le i\le n} m_i^1 - m^1/n) \le \frac{1}{\nu}(m^1 - m^1/n) = \frac{n-1}{n\nu} m^1 \le \frac{\bar{\rho} d(n-1)\bar{l}_1}{n\nu}$$

Similarly, we can get an upper bound of Δt_q as follows

$$\Delta t_q \le \frac{\bar{\rho} d(n-1)\bar{l}_q}{n\nu}.$$

Since $S = \bigcup_{k=1}^{q} A_k$ and $\Delta T = \sum_{k=1}^{q} \Delta t_k$, the conclusion follows. □

3.4 Simulation Results

In this section, we first provide numerical examples to verify our coverage algorithm DSCA. At first we consider 5 agents in a rectangular region D with parameters given as follow: $l = 6$, $h = 8$, $d = 1$ (that is, $q = 8$), $\kappa = 4$, $\underline{\rho} = 1$, $\bar{\rho} = 10$ and

$$\rho(x, y) = 5.5 + 4.5 \sin(x + y).$$

Moreover, 5 agents with $\nu = 5$ and the sampling period $T_s = 10^{-3}$ are used to sweep the region. The sweep process of implementing the DSCA is shown in Fig. 3.4. At the initial time, the first stripe has been divided into 5 sub-stripes with equal workload without loss of generality (i.e., $H_1^b = 0$). Then the sweeping and partitioning operations are carried out simultaneously, where the vertical lines in Fig. 3.4 describe the final partition marks (that is, the boundaries between the sub-stripes) on the corresponding stripe. The total time to cover the whole region by the five cooperative agents is 10.56 in the simulation. Since the optimal coverage time is 10.51, the extra time to sweep D for the DSCA is 0.05, which is less than the estimation 0.72 according to (3.12).

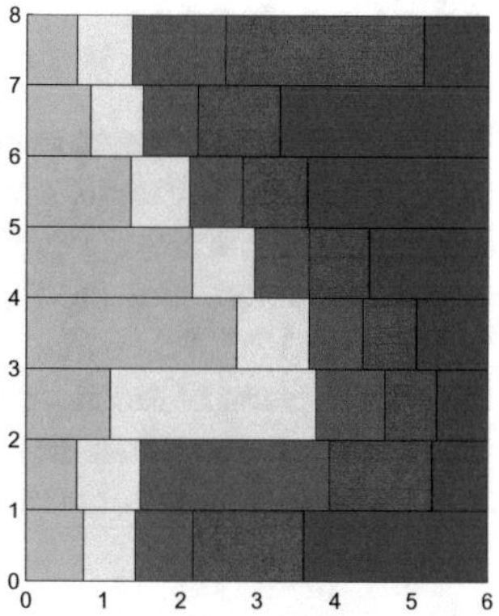

Fig. 3.4 (color online) Sweep process of 5 agents using DSCA in a rectangular region. Colored parts show what has been swept. Sub-stripes of the same color are covered by the same agent

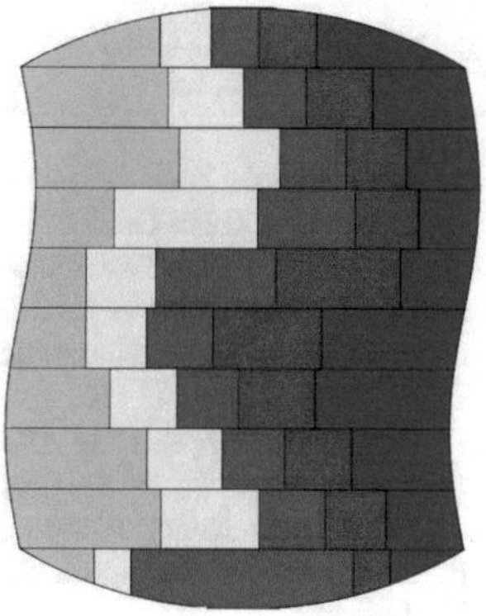

Fig. 3.5 Sweep coverage of 5 agents for region S

Next, we consider the coverage of a general bounded region S (shown in Fig. 3.5), which is enclosed by four curves:

$$y = 5 - \sqrt{10x - x^2}, \quad y = 5 + \sqrt{10x - x^2}$$

and

$$x = 0.2 \sin \frac{\pi(y-5)}{4}, \quad x = 0.2 \sin \frac{\pi(y-5)}{4} + 8$$

The workload distribution function is given by

$$\rho(x, y) = 4 + 2\sin(x + y)$$

with $\underline{\rho} = 2$ and $\bar{\rho} = 6$. Other parameters are the same as above. Since the total coverage time and optimal coverage time are 10.21 and 9.03, respectively, the extra time to sweep the given region is 1.18, less than the upper bound estimation 11.54 given by (3.13) with $H_2^b = 0.01$, $\tau_d = 3$ and $\sigma_d = 1$.

3.5 Conclusions

In this chapter, a cooperative coverage algorithm was proposed to handle the dynamic sweep problem for multi-agent systems in a region with uncertain workload. The decentralized coverage algorithm was designed to achieve the sweep coverage of the given region, and the theoretical analysis was also conducted to estimate an upper bound of the coverage time spent more than the optimal time. Moreover, the effectiveness of the algorithms was verified by numerical simulations. However, many problems remain to be solved, including adaptive sweeping along with partition in a parametric environment and the influence of different communication links between agents.

References

1. Anderson, B., Yu, B., Fidan, B., Hendrickx, J.: Rigid graph control architectures for autonomous formations. IEEE Control Syst. Mag. **28**(6), 48–63 (2008)
2. Nancy, A.: Lynch Distributed Algorithms. Morgan Kaufmann (1997)
3. Hong, Y., Hu, J., Gao, L.: Tracking control for multi-agent consensus with an active leader and variable topology. Automatica **42**(9), 1177–1182 (2006)
4. Chen, Z., Zhang, H.: No-beacon collective circular motion of jointly connected multi-agents. Automatica **47**(9), 1929–1937 (2011)
5. Cortés, J., Martínez, S., Karatas, T., Bullo, F.: Coverage control for mobile sensing network. IEEE Trans. Robot. Autom. **20**(2), 243–255 (2004)
6. Cassandras, C.G., Li, W.: Sensor networks and cooperative control. Eur. J. Control. **11**(4–5), 436–463 (2005)
7. Hussein, I.I., Stipanović, D.: Effective coverage control for mobile sensor networks with guaranteed collision avoidance. IEEE Trans. Control Syst. Technol. **15**(4), 642–657 (2007)
8. Wang, Y., Hussein, I.I.: Awareness coverage control over large-scale domains with intermittent communications. IEEE Trans. Autom. Control **55**(8), 1850–1859 (2010)
9. Cheng, T., Savkin, A.: A distributed self-deployment algorithm for the coverage of mobile wireless sensor networks. IEEE Commun. Lett. **13**(11), 877–879 (2009)
10. Gage, D.: Command control for many-robot systems. In: Proceedings of the 19th Annual AUVS Teachnical Symposium, pp. 22–24. Huntsville, Alabama (1992)
11. Cheng, T., Savkin, A.: Decentralized coordinated control of a vehicle network for deployment in sweep coverage. In: Proceedings of the IEEE International Conference on Control and Automation, pp. 275–279. Christchurch, New Zealand (2009)
12. Hu, X., Huang, Y., Cheng, D.: Optimal design of grazing behavior for multi-agent robots. In: Cheng, D., Sun, Y., Shen, T., Ohmori, H., (eds.), Advanced Robust and Adaptive Control, pp. 71–84. Tsinghua University Press, Beiijng (2005)
13. Choset, H.: Coverage of known spaces: The boustrophedon cellular decomposition. Auton. Robot. **9**(3), 247–253 (2000)
14. Butler, Z., Rizzi, A., Hollis, R.: Complete distributed coverage of rectilinear environments. In: Proceedings of the Workshop on the Algorithmic Foundations of Robotics (2000)
15. Min, T., Yin, H.: A decentralized approach for cooperative sweeping by multiple mobile robots. In: Proceedings of the IEEE/RSJ International Conference on Intelligent Robots and Systems, 13-17, vol. 1, pp. 380–385 (1998)

16. Wagner, I., Lindenbaum, M., Bruckstein, A.: Distributed covering by ant-robots using evaporating traces. IEEE Trans. Robot. Autom. **15**(5), 918–933 (1999)
17. Choset, H.: Coverage for robotics-A survey of recent results. Ann. Math. Artif. Intell. **31**(1–4), 113–126 (2001)
18. Demmel, J.: Applied Numerical Linear Algebra. SIAM, Philadelphia (1997)

Chapter 4
Adaptive Cooperative Coverage Algorithm with Online Learning Strategies

4.1 Introduction

Recent years have witnessed the increasing research attention on multi-agent systems, with many applications in distributed information collection and cooperative control [1–3]. Multi-agent coverage is one of the important cooperative problems, which is related to exploration, monitoring, and surveillance. Up to now, many formulations have been proposed for different coverage problems [4–6]. The uncertainty in the environment causes many troubles in the coverage design. As a result, many results were obtained for the region coverage based on adaptive control. For example, an adaptive algorithm was used to learn environment information online in [7], while the problem of positioning a team of robots in the non-convex region for a surveillance task was addressed by using the cognitive-based adaptive optimization algorithm in [8].

On the other hand, sweep coverage is an important coverage type that can be found in many applications [9]. Different from many coverage problems, sweep coverage is a dynamic coverage problem. The agents cannot cover a given region at one time. Instead, the group of agents have to collaborate each other and move in the given region to detect targets of interest or complete workload. Reference [10] proposed a formation-based method sweep coverage strategy. It is difficult in that all the agents have to cooperate in order to optimize the operation time, especially when there are environmental uncertainties. There are few results on the research. Reference [11] discussed the dynamic sweep coverage with bounded uncertainties in the covered region.

The objective of this chapter is to propose an adaptive dynamic sweep coverage to deal with the parametric uncertainties in the considered region. Note that the existing adaptive coverage algorithms were not about dynamic sweeping, and the design given in [11] was for bounded uncertainties. Here for parametric uncertainties, we provide an adaptive coverage algorithm by considering both partition (to handle the uncertainties) and sweep (to complete the coverage). Here an adaptive control technique is adopted to deal with the unknown dynamical environment for

C. Zhai et al., *Cooperative Coverage Control of Multi-Agent Systems and its Applications*, Studies in Systems, Decision and Control 408,
https://doi.org/10.1007/978-981-16-7625-3_4

the estimation on the difference between the actual coverage time and the optimal time, though it may not achieve the sweep coverage within optimal operation time due to the uncertainty. Moreover, the uncertain region is unbounded, different from the region in [11, 12].

The remainder of the chapter is organized as follows. A formulation of sweep coverage in uncertain environment is presented in Sect. 4.2. Then, a decentralized sweep coverage algorithm and the estimation of the extra time are shown in Sect. 4.3. Finally, conclusions are given in the last section.

4.2 Problem Formulation

In this section, we consider the adaptive design to sweep the region with unknown workload distribution by a MAS. In the coverage control, the whole region is usually divided into several subregions, and each agent is responsible for the coverage workload on its own subregion to complete the sweep coverage in a cooperative way. If there is no uncertainty and workload distribution can be known for all the agents in advance, the optimal strategy to complete the coverage can be carried out by partitioning the whole region into subregions with equal workload for each agent. However, due to the uncertainties of workload distribution, the optimal partition cannot be provided, and we have to design new algorithms to try our best for the sweep coverage.

In our problem, we consider an unbounded rectangular region D with width l_a (see Fig. 4.1). Different from the bounded region discussed in [11], we will use adaptive control scheme to learn the uncertain environment parameter (as time t tends to infinity) and then estimate the error bound for the extra time spent in addition to that spent by the optimal policy without any uncertainties. All the agents line up at the bottom of D, and they will move upwards by sweeping the whole region. Suppose each agent has its actuation range with diameter d. When it sweeps when it moves, it will clean up a stripe with width d. For simplicity, the whole region is partitioned into stripes (with the kth stripe denoted as A_k for $k = 1, 2, \cdots$) with length l_a and width d, and the agents sweep these stripes one by one to complete the whole region coverage. To shorten the whole time for coverage, each stripe is partitioned into sub-stripes for each agent so that each agent has the same workload. Then each agent sweeps its own sub-stripe and simultaneously partitions the next stripes for sweep. The workload distribution on D is denoted by $\rho(x, y) > 0$, which is continuous and bounded. Suppose ρ can be written as

$$\rho(x, y) = K(x, y)^T a, \tag{4.1}$$

where $a \in R_+^m$ is an uncertain parameter vector, and $K(x, y)$ with each element positive, is available to each agent. The following assumption is useful for the adaptive design, which means the environment contains much information. In fact, it is corre-

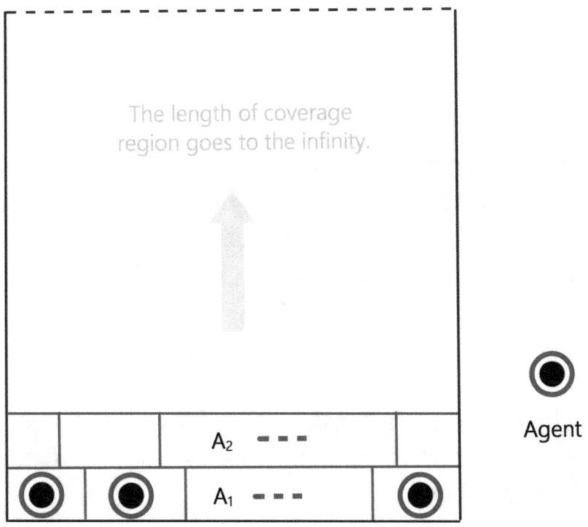

Fig. 4.1 Sweep coverage of multiple mobile agents on the region D

sponding to the persistence condition to guarantee the convergence of the estimation of uncertain parameters [13].

Assumption 4.1 There exists a constant $\bar{c} > 0$ such that for any $x \in (0, l_a)$

$$\Omega_k(x)\Omega_k^T(x) \geq \bar{c}I, \quad k \in \mathbb{Z}^+$$

where I is the identity matrix and

$$\Omega_k(x) = \int_{y_{k-1}}^{y_{k-1}+d} K(x, y)dy$$

is the distribution of x on the kth stripe, with $y_0 = 0$ and $y_k = y_{k-1} + d$.

For simplicity, we assume the initial stripe is partitioned into sub-stripes with equal workload for each agent. Moreover, each agent has the same sweeping capability v. Obviously, the completion time of sweeping each sub-stripe only depends on v and workload on the sub-stripe without considering the area and shape of sub-stripe. Moreover, each agent can detect the workload around its position and estimate workload at other positions according to prior information on workload distribution. When each agent sweeps its own sub-stripe, the partition on the next stripe is conducted according to the proposed adaptive partition algorithm. Once all the agents finish sweeping their own sub-stripes, the partition operation stops. Then, all the agents move to the newly-partitioned stripes and repeat sweep operations until the whole sweeping task is completed. Additionally, we assume each agent always gets the information of the neighbors in adjacent sub-stripes.

Let $\hat{a}_i$ denote the estimation of the parameter vector by agent i, where $i \in E_n = \{1, 2, ..., n\}$ and the parameter error $\tilde{a}_i = \hat{a}_i - a$. Then the workload on the sub-stripe of the stripe A_k between partition mark $i - 1$ and partition mark i estimated by agent

$j \in \{i-1, i\}$ is given by

$$\hat{m}_i^j = \int_{x_{i-1}}^{x_i} \Omega_k^T(\tau)\hat{a}_j d\tau$$

where x_{i-1} and x_i denote the positions of the partition marks $i-1$ and i, respectively ($x_0 = 0$ and $x_n = l_a$). For simplicity, A_k is regarded as a line segment with distribution density $\psi(x) = \Omega_k^T(x)a$ by integrating along the vertical axis. In addition, we assume that agent i can always detect $\psi(x_i)$. Here, the update law for $\hat{a}_i$ is designed as

$$\dot{\hat{a}}_i = -\mu\Omega_k(x_i)(\Omega_k^T(x_i)\hat{a}_i - \psi(x_i)) \tag{4.2}$$

where $\mu > 0$ is a constant gain. Next, we present the dynamic partition on the stripe A_k and the corresponding control law. The mathematical model of partition mark i is denoted by

$$\dot{x}_i = u_i, \tag{4.3}$$

where u_i is the partition control for partition mark i:

$$u_i = \kappa(\hat{m}_{i+1}^i - \hat{m}_i^i), \tag{4.4}$$

for a given constant $\kappa > 0$.

4.3 Technical Analysis

In this section, we give the theoretical analysis for the distributed sweep coverage algorithm and estimate the extra coverage time due to uncertainties. First, we provide the following lemma for the convergence analysis of the update law (4.2) on the stripe. Define

$$F = \sum_{i=1}^{n-1} \tilde{a}_i^T \tilde{a}_i.$$

then we have the following lemma.

Lemma 4.1 *With Assumption 4.1, F is exponentially convergent under the update law (4.2).*

Proof The derivative of F along the trajectories of (4.2) satisfies

$$\begin{aligned}\dot{F} &= 2\sum_{i=1}^{n-1} \tilde{a}_i^T \dot{\hat{a}}_i \\ &= -2\mu \sum_{i=1}^{n-1} \tilde{a}_i^T \Omega_k(x_i)(\Omega_k^T(x_i)\hat{a}_i - \psi(x_i)) \\ &= -2\mu \sum_{i=1}^{n-1} \tilde{a}_i^T \Omega_k(x_i)\Omega_k^T(x_i)\tilde{a}_i\end{aligned}$$

According to Assumption 4.1, we have

$$\mu \sum_{i=1}^{n-1} \tilde{a}_i^T \Omega_k(x_i)\Omega_k^T(x_i)\tilde{a}_i \geq \mu\bar{c} \sum_{i=1}^{n-1} \tilde{a}_i^T \tilde{a}_i$$

Hence, we have

$$\dot{F} \leq -2\mu\bar{c} \sum_{i=1}^{n-1} \tilde{a}_i^T \tilde{a}_i = -2\mu\bar{c}F$$

By solving the above differential inequality, we conclude that

$$F(t) \leq F(0)e^{-2\mu\bar{c}t}. \tag{4.5}$$

Thus, we completes the proof of this lemma. □

Remark 4.1 From Lemma 4.1, we can easily find that

$$||\tilde{a}_i(t)|| \leq \sqrt{F(0)}e^{-\mu\bar{c}t}.$$

To measure the uniformity of workload partition on A_k, we define

$$H_k = \sum_{i=1}^{n} (m_i - \bar{m})^2,$$

where

$$m_i = \int_{x_{i-1}}^{x_i} \psi(\tau)d\tau, \quad \bar{m} = \frac{1}{n}\int_0^{l_a} \psi(\tau)d\tau$$

and it can be rewritten as

$$H_k = \sum_{i=1}^{n} m_i^2 - n\bar{m}^2.$$

Obviously, the smaller H_k is, the more well-proportioned workload partition we will get. In addition, H_k^a and H_k^b denote the uniformity of workload partition at initial and final partition positions on A_k, respectively. The following lemma can be found in [11].

Lemma 4.2 *On the stripe A_k, the following inequality holds:*

$$\frac{\underline{\lambda}}{n} H_k \leq \sum_{i=1}^{n-1} (m_{i+1} - m_i)^2 \leq \bar{\lambda} H_k.$$

where $\underline{\lambda}$ and $\bar{\lambda}$ are the minimum and maximum eigenvalues of the positive definite matrix Γ as follows:

$$\Gamma = \begin{pmatrix} 2\,0\,1 \,.\,.\, 1\,1\,2 \\ 0\,3\,0 \,.\,.\, 1\,1\,2 \\ 1\,0\,3 \,.\,.\, 1\,1\,2 \\ . \; . \; . \; . \; . \; . \; . \; . \; . \\ . \; . \; . \; . \; . \; . \; . \; . \; . \\ 1\,1\,1 \,.\,.\, 3\,0\,2 \\ 1\,1\,1 \,.\,.\, 0\,3\,1 \\ 2\,2\,2 \,.\,.\, 2\,1\,5 \end{pmatrix} \in R^{(n-1)\times(n-1)},$$

for $n > 2$ or $\Gamma = 4$ for $n = 2$.

For simplicity, we introduce two parameters as follows

$$\underline{\psi}_k = \inf_{0 \leq x \leq l_a} \sum_{j=1}^{m} \Omega_k^j(x) a(j)$$

and

$$\bar{\psi}_k = \sup_{0 \leq x \leq l_a} \sum_{j=1}^{m} \Omega_k^j(x) a(j),$$

where $\Omega_k^j(x)$ and $a(j)$ represent the jth element of the vectors $\Omega_k(x)$ and a, respectively.

Lemma 4.3 *For the update law (4.2) and partition dynamics (4.3) with the control law (4.4), we have*

$$\lim_{t \to +\infty} |m_i - m_j| = 0, \quad \forall\, i, j \in E_n.$$

Proof The time derivative of H_k with respect to (4.3) is given by,

$$\begin{aligned}\dot{H}_k = & -2\kappa \sum_{i=1}^{n-1} (m_{i+1} - m_i)^2 \psi(x_i) \\ & + 2\kappa \sum_{i=1}^{n-1} (m_i - m_{i+1})(\tilde{m}_{i+1}^i - \tilde{m}_i^i)\psi(x_i)\end{aligned}$$

From Lemma 4.2, we have

$$\sum_{i=1}^{n-1} (m_{i+1} - m_i)^2 \psi(x_i) \geq \frac{\lambda}{n} H_k \psi(x_i) \geq \frac{\lambda \underline{\psi}_k}{n} H_k$$

and

$$\begin{aligned}& \sum_{i=1}^{n-1} (m_i - m_{i+1})(\tilde{m}_{i+1}^i - \tilde{m}_i^i)\psi(x_i) \\ & \leq \bar{\psi}_k (\sum_{i=1}^{n-1} (m_i - m_{i+1})^2)^{\frac{1}{2}} (\sum_{i=1}^{n-1} (\tilde{m}_{i+1}^i - \tilde{m}_i^i)^2)^{\frac{1}{2}} \\ & \leq \bar{\psi}_k (\bar{\lambda} H_k)^{\frac{1}{2}} (\sum_{i=1}^{n-1} (\omega_k^T \tilde{a}_i)^2)^{\frac{1}{2}} \\ & \leq \bar{\psi}_k (\bar{\lambda} H_k)^{\frac{1}{2}} (\sum_{i=1}^{n-1} ||\omega_k||^2 ||\tilde{a}_i||^2)^{\frac{1}{2}}\end{aligned}$$

where $\|\cdot\|$ denotes the Euclidean norm and ω_k is defined as

$$\omega_k = \int_{x_i}^{x_{i+1}} \Omega_k(\tau) d\tau - \int_{x_{i-1}}^{x_i} \Omega_k(\tau) d\tau.$$

Considering that

$$||\omega_k||^2 \leq || \int_0^{l_a} \Omega_k(\tau) d\tau ||^2,$$

we have

$$\begin{aligned}& \sum_{i=1}^{n-1} (m_i - m_{i+1})(\tilde{m}_{i+1}^i - \tilde{m}_i^i)\psi(x_i) \\ & \leq \bar{\psi}_k (\bar{\lambda} H_k)^{\frac{1}{2}} (\sum_{i=1}^{n-1} ||\tilde{a}_i||^2)^{\frac{1}{2}} || \int_0^{l_a} \Omega_k(\tau) d\tau || \\ & = \bar{\psi}_k (\bar{\lambda} H_k)^{\frac{1}{2}} F^{\frac{1}{2}} || \int_0^{l_a} \Omega_k(\tau) d\tau ||.\end{aligned}$$

Therefore,

$$\dot{H}_k \leq -\frac{2\kappa\underline{\lambda\psi}_k}{n}H_k + 2\kappa\bar{\psi}_k(\bar{\lambda}H_k F)^{\frac{1}{2}}||\int_0^{l_a}\Omega_k(\tau)d\tau|| \tag{4.6}$$

To obtain a linear differential inequality, we take $G_k = \sqrt{H_k}$ (Likewise, we define $G_k^a = \sqrt{H_k^a}$ and $G_k^b = \sqrt{H_k^b}$). Since

$$\dot{G}_k = \dot{H}_k/2\sqrt{H_k}, \quad H_k \neq 0,$$

we see that

$$\dot{G}_k \leq -\frac{\kappa\underline{\lambda\psi}_k}{n}G_k + \kappa\bar{\psi}_k\sqrt{\bar{\lambda}F}||\int_0^{l_a}\Omega_k(\tau)d\tau||. \tag{4.7}$$

When $H_k = G_k = 0$, it can be shown that the inequality still holds. Hence, solving the differential inequality (4.7) yields

$$\begin{aligned} G_k(t) &\leq G_k(0)e^{-\frac{\kappa\underline{\lambda\psi}_k}{n}t} + \epsilon_k\int_0^t e^{\frac{\kappa\underline{\lambda\psi}_k}{n}(\tau-t)}\sqrt{F(\tau)}d\tau \\ &\leq (G_k(0) + \epsilon_k\int_0^t e^{\frac{\kappa\underline{\lambda\psi}_k}{n}\tau}\sqrt{F(\tau)}d\tau)e^{-\frac{\kappa\underline{\lambda\psi}_k}{n}t} \end{aligned} \tag{4.8}$$

where

$$\epsilon_k = \kappa\bar{\psi}_k\sqrt{\bar{\lambda}}||\int_0^{l_a}\Omega_k(\tau)d\tau||. \tag{4.9}$$

From (4.5) in Lemma 4.1, we have

$$\sqrt{F(\tau)} \leq \sqrt{F(0)}e^{-\mu\bar{c}\tau}$$

and

$$\int_0^t e^{\frac{\kappa\underline{\lambda\psi}_k}{n}\tau}\sqrt{F(\tau)}d\tau \leq \sqrt{F(0)}\int_0^t e^{(\frac{\kappa\underline{\lambda\psi}_k}{n}-\mu\bar{c})\tau}d\tau.$$

It follows from (4.8) that

$$\begin{aligned} G_k(t) &\leq \left(G_k(0) + \epsilon_k\sqrt{F(0)}\int_0^t e^{(\frac{\kappa\underline{\lambda\psi}_k}{n}-\mu\bar{c})\tau}d\tau\right)e^{-\frac{\kappa\underline{\lambda\psi}_k}{n}t} \\ &= (G_k(0) - \frac{\epsilon_k\sqrt{F(0)}}{\frac{\kappa\underline{\lambda\psi}_k}{n}-\mu\bar{c}})e^{-\frac{\kappa\underline{\lambda\psi}_k}{n}t} + \frac{\epsilon_k\sqrt{F(0)}}{\frac{\kappa\underline{\lambda\psi}_k}{n}-\mu\bar{c}}e^{-\mu\bar{c}t} \end{aligned}$$

when

$$\frac{\kappa\underline{\lambda\psi}_k}{n} - \mu\bar{c} \neq 0$$

and

$$G_k(t) \leq (G_k(0) + t\epsilon_k\sqrt{F(0)})e^{-\frac{\kappa\lambda\underline{\psi}_k}{n}t}$$

for

$$\frac{\kappa\lambda\underline{\psi}_k}{n} - \mu\bar{c} = 0.$$

Thus, we get

$$\lim_{t\to+\infty} G_k(t) = \lim_{t\to+\infty} \sqrt{H_k(t)} = 0,$$

which implies the conclusion. □

Note that the boundaries between sub-stripes on A_k are exactly the final positions of the partition marks on the stripe. Clearly, the initial positions of partition marks on A_{k+1} is the final partition position, that is, the boundaries between sub-stripes on A_k, respectively. Thus, we can get the result as follows.

Lemma 4.4 *H^a_{k+1} and H^b_k satisfy the following inequality*

$$H^a_{k+1} \leq \alpha_k^2 H^b_k + \beta_k \frac{l_a^2}{n} \tag{4.10}$$

where

$$\alpha_k = \frac{\bar{\psi}_{k+1}}{\underline{\psi}_k}$$

and

$$\beta_k = \frac{\bar{\psi}_k^2 \bar{\psi}_{k+1}^2}{\underline{\psi}_k^2} - \underline{\psi}_{k+1}^2.$$

The proof is nearly the same as that of Lemma 3.3 in [11] and thus omitted. To prove the convergence of the extra coverage time, we make the following assumption.

Assumption 4.2 There are $k^* \in \mathbb{Z}^+$ and $\bar{\alpha} > 0$ such that for any $k \geq k^*$

$$H^a_{k+1} \leq \bar{\alpha}^2 H^b_k$$

holds with

$$\bar{\alpha} < e^{\frac{\kappa\lambda\underline{\psi}}{n}\underline{t}},$$

where

$$\underline{\psi} = \inf_{k\in\mathbb{Z}^+} \underline{\psi}_k, \quad \underline{t} = \frac{l_a\underline{\psi}}{n\nu}$$

Remark 4.2 Actually, this assumption is not strong compared with the inequality (4.10), since $\bar{\alpha}$ can be large enough by adjusting the parameter κ. In addition,

Table 4.1 Decentralized sweep coverage algorithm

Goal: sweep the unbounded region D
$k = 1$
while workload on D is not completed **do**
if workload on A_k is not completed **then**
Each agent sweeps its own sub-stripe on A_k
and executes the adaptive partition algorithm on A_{k+1}.
else
Each agent stops the adaptive partition operation
and moves to its corresponding sub-stripe on A_{k+1}.
$k = k + 1$
end if
end while

Assumption 4.2 holds naturally if $\rho(x, y)$ can be written as

$$\rho(x, y) = f(x)g(y),$$

where $f(x)$ and $g(y)$ are functions of x and y, respectively. Essentially, this assumption guarantees the monotonic decrease of H_k^b when $k \geq k^*$.

Our aim is to complete the coverage task as soon as possible. If all the environmental information is known in advance, we can conduct the optimal partition offline and then achieve the optimal coverage time. However, because of the uncertainties, we have to do partition online along with the sweep motion, which requires more time to complete the whole sweep coverage. Thus, we estimate the extra time to cover D besides the optimal coverage time using the DECENTRALIZED SWEEP COVERAGE ALGORITHM shown in Table 4.1.

Theorem 4.1 *Suppose the first stripe is partitioned with equal workload and*

$$\frac{\kappa\underline{\lambda}\underline{\psi}_k}{n} - \mu\bar{c} < 0.$$

Both Assumptions 4.1 and 4.2 hold. For the DECENTRALIZED SWEEP COVERAGE ALGORITHM, *the extra time to sweep the unbounded region D spent more than the optimal coverage time is bounded by*

$$
\begin{aligned}
\Delta T \le\ & \frac{1}{\nu}\sum_{k=1}^{k^*-1}\sum_{i=1}^{k}\left(\frac{\gamma_i}{\alpha_i}\prod_{j=i}^{k}\frac{\alpha_j}{e^{\frac{\kappa\lambda}{n}t_n^j\underline{\psi}_{j+1}}}\right) \\
&+\frac{1}{\nu(1-\bar{\bar{\alpha}})}\left(\bar{\bar{\alpha}}\sum_{i=1}^{k^*-1}\left(\frac{\gamma_i}{\alpha_i}\prod_{j=i}^{k^*-1}\frac{\alpha_j}{e^{\frac{\kappa\lambda}{n}t_n^j\underline{\psi}_{j+1}}}\right)+\frac{\bar{\epsilon}e^{-k^*\underline{t}}}{e^{\underline{t}}-1}\right)
\end{aligned} \tag{4.11}
$$

where

$$
\bar{\bar{\alpha}} = \bar{\alpha}e^{-\frac{\kappa\lambda\underline{\psi}}{n}\underline{t}}, \quad t_n^j = \frac{l_a\underline{\psi}_j}{n\nu}
$$

and

$$
\bar{\epsilon} = \sup_{k\in\mathbb{Z}^+}\frac{\epsilon_k\sqrt{F(0)}}{\mu\bar{c}-\frac{\kappa\lambda\underline{\psi}_k}{n}}
$$

and

$$
\gamma_i = l_a\sqrt{\frac{\beta_i}{n}}+\epsilon_{i+1}\sigma_{i+1}\sqrt{F(0)}, \quad \sigma_i = \frac{1}{\mu\bar{c}-\frac{\kappa\lambda\underline{\psi}_i}{n}}
$$

and ϵ_i is given in (4.9).

Proof Since all the agents sweep their respective sub-stripes simultaneously, the time of completing sweeping the stripe depends on the sub-stripe with the most workload. Firstly, we estimate the difference between the most workload and average workload on the subregion of stripe A_k, which is given by

$$
\max_{1\le i\le n} m_i - \bar{m} \le \left(\sum_{i=1}^{n}(m_i-\bar{m})^2\right)^{\frac{1}{2}} = \sqrt{H_k^b} = G_k^b.
$$

Then the extra time to sweep stripe A_k compared to the optimal partition satisfies

$$
\Delta t_k = \frac{1}{\nu}(\max_{1\le i\le n} m_i - \bar{m}) \le \frac{G_k^b}{\nu}.
$$

From (4.8) in Lemma 4.3 and $\frac{\kappa\lambda\underline{\psi}_k}{n} - \mu\bar{c} < 0$, we have

$$
\begin{aligned}
G_k(t) &\le (G_k(0)+\epsilon_k\int_0^t e^{\frac{\kappa\lambda\underline{\psi}_k}{n}\tau}\sqrt{F(\tau)}d\tau)e^{-\frac{\kappa\lambda\underline{\psi}_k}{n}t} \\
&\le (G_k(0)+\epsilon_k\sigma_k\sqrt{F(0)})e^{-\frac{\kappa\lambda\underline{\psi}_k}{n}t}
\end{aligned}
$$

Note that the time used to partition stripe A_{k+1} by n agents is the time to sweep stripe A_k with the lower bound $t_n^k = \frac{l_a\underline{\psi}_k}{n\nu}$. Therefore, we get

$$\begin{aligned} G_{k+1}^{b} &\leq G_{k+1}(t_n^k) \\ &\leq (G_{k+1}^{a} + \epsilon_{k+1}\sigma_{k+1}\sqrt{F(0)})e^{-\frac{\kappa\lambda\underline{\psi}_{k+1}}{n}t_n^k} \end{aligned} \tag{4.12}$$

Moreover, it follows from (4.10) that

$$G_{k+1}^{a} \leq \alpha_k G_k^b + l_a\sqrt{\frac{\beta_k}{n}}. \tag{4.13}$$

Substituting (4.13) into (4.12), we get

$$G_{k+1}^{b} \leq (\alpha_k G_k^b + l_a\sqrt{\frac{\beta_k}{n}} + \epsilon_{k+1}\sigma_{k+1}\sqrt{F(0)})e^{-\frac{\kappa\lambda\underline{\psi}_{k+1}}{n}t_n^k} \tag{4.14}$$

Since the first stripe is partitioned with equal workload, the extra time to sweep A_1 is $\Delta t_1 = 0$. Combining (4.14) with the initial value $G_1^b = 0$ yields,

$$G_{k+1}^{b} \leq \sum_{i=1}^{k}\left(\gamma_i \prod_{j=i+1}^{k} \alpha_j e^{-\frac{\kappa\lambda}{n}\sum_{s=i}^{k} t_n^s \underline{\psi}_{s+1}}\right).$$

Finally, since

$$\prod_{j=i+1}^{k} \alpha_j = \frac{1}{\alpha_i}\cdot\alpha_i \prod_{j=i+1}^{k} \alpha_j = \frac{1}{\alpha_i}\prod_{j=i}^{k}\alpha_j,$$

we obtain the upper bound of the extra time of sweeping the preceding k^* stripes as follows

$$\begin{aligned} \sum_{k=1}^{k^*} \Delta t_k &\leq \frac{1}{\nu}\sum_{k=1}^{k^*-1} G_{k+1}^{b} \\ &\leq \frac{1}{\nu}\sum_{k=1}^{k^*-1}\sum_{i=1}^{k}\left(\gamma_i \prod_{j=i+1}^{k} \alpha_j e^{-\frac{\kappa\lambda}{n}\sum_{s=i}^{k} t_n^s \underline{\psi}_{s+1}}\right) \\ &\leq \frac{1}{\nu}\sum_{k=1}^{k^*-1}\sum_{i=1}^{k}\left(\frac{\gamma_i}{\alpha_i}\prod_{j=i}^{k}\frac{\alpha_j}{e^{\frac{\kappa\lambda}{n}t_n^j\underline{\psi}_{j+1}}}\right) \end{aligned} \tag{4.15}$$

Moreover, solving the differential inequality (4.7) within the interval $[t_0, t]$ yields

$$
\begin{aligned}
G_k(t) &\le G_k(t_0)e^{-\frac{\kappa\underline{\lambda}\underline{\psi}_k}{n}(t-t_0)} + \epsilon_k \int_{t_0}^{t} e^{\frac{\kappa\underline{\lambda}\underline{\psi}_k}{n}(\tau-t)}\sqrt{F(\tau)}d\tau \\
&\le G_k(t_0)e^{-\frac{\kappa\underline{\lambda}\underline{\psi}_k}{n}(t-t_0)} + \frac{\epsilon_k\sqrt{F(t_0)}}{\mu\bar{c} - \frac{\kappa\underline{\lambda}\underline{\psi}_k}{n}} e^{-\frac{\kappa\underline{\lambda}\underline{\psi}_k}{n}(t-t_0)} \\
&\le G_k(t_0)e^{-\frac{\kappa\underline{\lambda}\underline{\psi}_k}{n}(t-t_0)} + \frac{\epsilon_k\sqrt{F(0)}}{\mu\bar{c} - \frac{\kappa\underline{\lambda}\underline{\psi}_k}{n}} e^{-\frac{\kappa\underline{\lambda}\underline{\psi}_k}{n}t} \\
&\le G_k(t_0)e^{-\frac{\kappa\underline{\lambda}\underline{\underline{\psi}}}{n}(t-t_0)} + \bar{\epsilon}e^{-\frac{\kappa\underline{\lambda}\underline{\underline{\psi}}}{n}t}.
\end{aligned}
$$

Thus, we have

$$
G_k^b \le G_k^a e^{-\frac{\kappa\underline{\lambda}\underline{\underline{\psi}}}{n}\underline{t}} + \bar{\epsilon}e^{-\frac{\kappa\underline{\lambda}\underline{\underline{\psi}}}{n}k\underline{t}}.
$$

It follows from Assumption 4.2 that

$$
G_{k+1}^b \le \bar{\bar{\alpha}}G_k^b + \bar{\epsilon}e^{-\frac{\kappa\underline{\lambda}\underline{\underline{\psi}}}{n}(k+1)\underline{t}}
$$

when $k \ge k^*$. Note that

$$
\begin{aligned}
\sum_{k=k^*+1}^{s} \Delta t_k &= \frac{1}{\nu}\sum_{k=k^*+1}^{s} G_k^b \\
&\le \frac{1}{\nu}\left[G_{k^*}^b \sum_{i=1}^{s-k^*}\bar{\bar{\alpha}}^i + \bar{\epsilon}\left(\sum_{i=k^*+1}^{s} e^{-i\underline{t}}\right)\sum_{i=0}^{s-k^*-1}\bar{\bar{\alpha}}^i\right].
\end{aligned}
$$

Since

$$
\bar{\bar{\alpha}} = \bar{\alpha}e^{-\frac{\kappa\underline{\lambda}\underline{\underline{\psi}}}{n}\underline{t}} < 1,
$$

we get

$$
\begin{aligned}
&\sum_{k=k^*+1}^{\infty} \Delta t_k \\
&= \lim_{s\to+\infty}\sum_{k=k^*+1}^{s} \Delta t_k \\
&\le \frac{1}{\nu(1-\bar{\bar{\alpha}})}\left(\bar{\bar{\alpha}}G_{k^*}^b + \frac{\bar{\epsilon}e^{-k^*\underline{t}}}{e^{\underline{t}}-1}\right) \\
&\le \frac{1}{\nu(1-\bar{\bar{\alpha}})}\left(\bar{\bar{\alpha}}\sum_{i=1}^{k^*-1}\left(\frac{\gamma_i}{\alpha_i}\prod_{j=i}^{k^*-1}\frac{\alpha_j}{e^{\frac{\kappa\underline{\lambda}}{n}t_n^j\underline{\psi}_{j+1}}}\right) + \frac{\bar{\epsilon}e^{-k^*\underline{t}}}{e^{\underline{t}}-1}\right)
\end{aligned}
\tag{4.16}
$$

Clearly, we have

$$\Delta T = \sum_{k=1}^{\infty} \Delta t_k = \sum_{k=1}^{k^*} \Delta t_k + \sum_{k=k^*+1}^{\infty} \Delta t_k.$$

Combining (4.15) and (4.16), we get the upper bound of ΔT. □

Note that Theorem 4.1 shows that the extra time spent to cover the region due to the parameter uncertainty is bounded by (4.11) after learning online based on update law (4.2), though there are errors after moving from a stripe to the next one. The upper bound given in (4.11) is in fact based on the real value a, which is not known in the beginning. However, if the domain D_a of a is known and bounded, then we can easily select $\hat{a}$ in D_a and estimate ΔT based on our analysis.

Remark 4.3 The condition

$$\frac{\kappa \underline{\lambda} \underline{\psi}_k}{n} - \mu \bar{c} < 0$$

guarantees that the convergent rate of F is larger than that of G_k.

4.4 Conclusions

In this chapter, an adaptive algorithm was proposed to deal with the sweep coverage problem for multiple agent networks in a region with parametric uncertainties. The analysis was presented to achieve the complete coverage of the given region and estimate the upper bound of the coverage time spent more than the optimal time. Moreover, convergence on the decentralized sweep coverage algorithm was carried out. Further investigation may include the relaxation of assumptions and the consideration of the sweeping dynamics of agents.

References

1. Zhang, H., Zhai, C., Chen, Z.: A general alignment repulsion algorithm for flocking of multi-agent systems. IEEE Trans. Autom. Contr. **56**(2), 430–435 (2011)
2. Hong, Y., Hu, J., Gao, L.: Tracking control for multi-agent consensus with an active leader and variable topology. Automatica **42**(9), 1177–1182 (2006)
3. Martínez, S., Cortés, J., Bullo, F.: Motion coordination with distributed information. IEEE Control Mag. **27**(4), 75–88 (2007)
4. Choset, H.: Coverage of known spaces: the boustrophedon cellular decomposition. Auton. Robots **9**(3), 247–253 (2000)
5. Du, Q., Faber, V., Gunzburger, M.: Centroidal Voronoi tesseuations: applications and algorithms. SIAM Rev. **41**(4), 637–676 (1999)
6. Cheng, T., Savkin, A.: A distributed self-deployment algorithm for the coverage of mobile wireless sensor networks. IEEE Commun. Lett. **13**(11), 877–879 (2009)
7. Schwager, M., Rus, D., Slotine, J.: Decentralized, adaptive control for coverage with networked robots. Int. J. Robot. Res. **28**(3), 357–375 (2009)

8. Renzaglia, A., Doitsidis, L., Martinelli, A., Kosmatopoulos, E.: Adaptive-based distributed cooperative multi-robot coverage. In: Proceedings of American Control Conference, pp. 486–473. San Francisco (2011)
9. Gage, D.: Command control for many-robot systems. In: Proceedings of the 19th Annual AUVS Teachnical Symposium, pp. 22–24. Huntsville, Alabama (1992)
10. Cheng, T., Savkin, A.: Decentralized coordinated control of a vehicle network for deployment in sweep coverage. In: Proceedings of the IEEE International Conference on Control and Automation, pp. 275–279. Christchurch, New Zealand (2009)
11. Zhai, C., Hong, Y.: Decentralized sweep coverage algorithm for uncertain region of multi-agent systems. In: Proceedings of American Control Conference. Montréal, Canada (2012), in press
12. Zhai, C., Hong, Y.: Decentralized algorithm for online workload partition of multi-agent systems. In: Proceedings of Chinese Control Conference. Yantai, China (2011), pp. 4920–4925
13. Ioannou, P., Sun, J.: Robust Adaptive Control. Prentice-Hall, Englewood Cliffs (1996)

Chapter 5
Distributed Sweep Coverage Algorithm Using Workload Memory

5.1 Introduction

The great progress in communication technologies makes it easy and low-cost to share mutual information and coordinate the joint actions among multiple agents. Thus, cooperative control of multi-agent systems has attracted much interest of researchers in various fields in the past decade. The coordination of multiple agents contributes to improving the efficiency and robustness of carrying out complicated tasks, such as leader tracking [1–3], flocking behavior [4, 5], boundary patrolling [6], persistent monitoring [7, 8] and region coverage [9], to name just a few.

As a type of coordination tasks, cooperative coverage of multi-agent systems refers to the path planning of a robot team to visit every point in the environment or the optimal deployment of sensor networks according to certain performance indexes. The approach of divide-and-conquer is widely applied in the region coverage of multi-agent systems [10–12]. Specifically, [11] presents a gradient descent algorithm to optimize a class of utility functions in the coverage region, where the centroidal Voronoi partition is adopted to allocate a subregion for each mobile sensor. Reference [12] extends the above work by proposing a distributed, adaptive coverage algorithm for nonholonomic mobile sensors. In multi-robot coordination, the coverage problem falls into three categories: blanket coverage, barrier coverage and sweep coverage [13]. Blanket coverage aims at deploying multiple agents in the given coverage region to maximize the probability of identifying the target [14]. Barrier coverage is used to protect the target in the given region and meanwhile maximize the detection rate of invaders overpassing the barrier that is formed by the agents [15]. As an important type of multi-robot coverage, sweep coverage can be used in many tasks, such as maintenance inspection [13], border patrolling [16] and environmental monitoring [17]. It is largely unexplored to investigate multi-agent sweep coverage in the uncertain environment, where the exact information on the coverage region is unknown in advance. Thus, the distributed control algorithm is developed to deal with the uncertainty in multi-agent sweep coverage. Sweep coverage can actually be regarded as a moving barrier, and it focuses on sweeping or

C. Zhai et al., *Cooperative Coverage Control of Multi-Agent Systems and its Applications*, Studies in Systems, Decision and Control 408,
https://doi.org/10.1007/978-981-16-7625-3_5

monitoring the given region by arriving at every point. For example, [16] investigates the sweep coverage of mobile sensors in a corridor environment without taking into account the workload distribution. To address this issue, [9] develops a sweep coverage algorithm by dividing the whole region into a series of stripes and then cooperatively completing the workload in each stripe in sequence. Nevertheless, it is not a fully distributed algorithm because agents have to shift stripes under the centralized command. Moreover, the partition error of workload accumulates rapidly as the number of stripes increases.

In this chapter, we investigate the sweep coverage problem of multi-agent systems in the uncertain environment, where the workload distribution is unknown in advance. The agents cooperatively partition the whole region using the trajectories of their partition bars while sweeping their respective subregions in a distributed manner. Compared with existing work, our approach achieves the fully distributed sweep coverage of multi-agent system in the uncertain environment by capitalizing on the workload memory, and the sweeping task can be fulfilled with guaranteed complete time in theory. Specifically, the main contributions of this work are listed as follows.

1. A distributed sweep coverage formulation of multi-agent systems is developed with workload memory, and an upper bound for the error between the actual sweeping time and the optimal time is estimated.
2. It is demonstrated that the multi-agent system is input-to-state stable with respect to the vertical speed of partition bars.
3. A sufficient condition is derived to avoid the collision of partition bars in the process of partitioning the coverage region.

The remainder of this chapter is organized as follows. Section 5.2 formulates the problem of distributed sweep coverage and proposes the sweep coverage algorithm of multi-agent systems. Sections 5.3 and 5.4 present theoretical results on the proposed sweep coverage algorithm, followed by numerical simulations in Sect. 5.5. Finally, we draw the conclusion and discuss the future direction in Sect. 5.6.

5.2 Problem Formulation

This section formulates the sweep coverage problem of multi-agent systems in the uncertain environment, where the distribution of workload (e.g., dust in the room, crops in the farmland, leaking oil in the sea, etc.) is unknown to the agents in advance. First of all, we present the formal description of multi-agent sweep coverage in the uncertain environment. Then the goal of multi-agent sweep coverage is provided and the multi-agent dynamics is designed mathematically. Finally, a distributed sweep coverage algorithm of multi-agent systems is proposed to fulfil the sweeping task in the uncertain region by means of divide and conquer.

The problem of multi-agent sweep coverage in the uncertain environment is described as follows: Consider the two dimensional coverage region Ω (see Fig. 5.1)

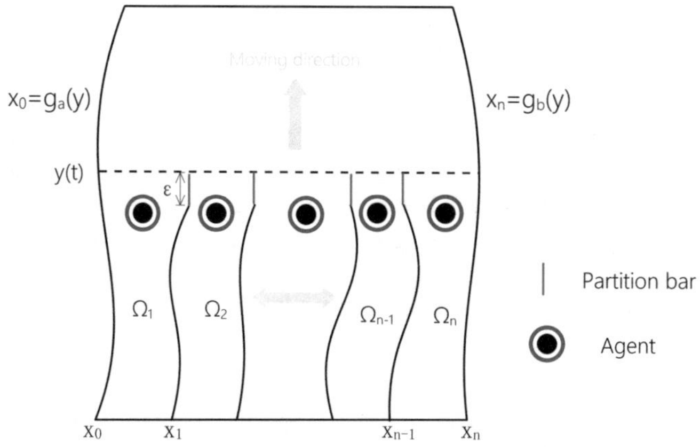

Fig. 5.1 Sweep coverage of the uncertain region Ω with smooth boundaries. The left and right boundaries of coverage region are described by $x = g_a(y)$ and $x = g_b(y)$, respectively. The blue line segments with the length ϵ denote the partition bars of agents, and their trajectories divide the partition region into n subregions (i.e., $\Omega_i(t), i \in \mathcal{I}_n$ at time t). The uncompleted part of the region Ω is marked in gray, while the completed part is marked in white

enclosed by two parallel horizontal lines with the distance l and two smooth boundaries described by $x = g_a(y)$ and $x = g_b(y)$, respectively. The distribution density function of workload is given by $\rho(x, y)$, satisfying $\rho(x, y) \in [\underline{\rho}, \bar{\rho}]$, where $\underline{\rho}$ and $\bar{\rho}$ are the lower and upper bounds of $\rho(x, y)$, respectively. To be precise, the uncertain environment is defined as the 2-tuple $(\Omega, \rho(x, y))$ with $\rho(x, y) \in [\underline{\rho}, \bar{\rho}]$, $\forall (x, y) \in \Omega$. There are n mobile agents responsible for sweeping the workload in the region Ω. Each agent can only detect the workload in its neighborhood and communicate with its neighbors to share the workload information. In addition, each agent is assumed to complete the workload on the coverage region at the constant sweeping rate of σ. In practice, σ depends on the sweeping performance of agents (e.g., sweeping robots, robot vacuums), and it is determined by the completed workload per unit time. In brief, the problem is how to design the multi-agent dynamics and allocate the workload for each agent so that the workload in the coverage region can be completed as soon as possible. This is a challenging problem because the global information on the workload distribution is unknown to the agents, which makes it impossible to equally divide the workload for each agent in advance. As a result, an online algorithm of multi-agent systems has to be developed to achieve the workload allocation for each agent by using the local workload information.

Suppose an online algorithm manages to divide the whole region Ω into n subregions (i.e., $\Omega = \bigcup_{i=1}^{n} \Omega_i$). Let T^* represent the optimal sweep time of multi-agent systems, and it can be achieved if the whole region is divided into n subregions with equal workload. In this way, all the agents can complete the workload in their own subregions simultaneously. Thus, the optimal sweep time T^* is given by $T^* = \bar{m}/\sigma$, where

$$\bar{m} = \frac{1}{n}\iint_{\Omega} \rho(x, y)dxdy = \frac{1}{n}\sum_{i=1}^{n}\iint_{\Omega_i} \rho(x, y)dxdy = \frac{1}{n}\sum_{i=1}^{n} m_i$$

denotes the average workload in each subregion, and m_i refers to the workload in the ith subregion Ω_i. In fact, T^* is normally unavailable due to the local information on workload distribution, and it is only regarded as the benchmark to assess the performance of a sweep coverage algorithm. Considering that an online algorithm normally leads to the unequal allocation of workload in subregions, the actual sweep time T mainly depends on the maximum workload in subregions. And it can be computed as:

$$T = \frac{1}{\sigma}\max_{i \in I_n} \iint_{\Omega_i} \rho(x, y)dxdy = \frac{1}{\sigma}\max_{i \in I_n} m_i, \quad I_n = \{1, 2, ..., n\},$$

Therefore, the error between the actual sweep time and the optimal sweep time ΔT is given by:

$$\Delta T = T - T^* = \frac{1}{\sigma}\left(\max_{i \in I_n} m_i - \bar{m}\right).$$

The goal of this paper is to design a real-time sweep coverage algorithm of multi-agent systems with local workload information to complete the workload in the region Ω and estimate the upper bound of time error ΔT.

Next, we propose the partition dynamics of multi-agent systems and then present the distributed sweep coverage algorithm in real time. To achieve the real-time workload partition, each agent is equipped with a partition bar with the length ϵ to allocate the workload in the whole region (see Fig. 5.1). Let (x_i, y_i), $i \in I_n$ refer to the coordinate position (i.e., the lower terminal point) of partition bar for agent i, then the dynamics of partition bars is designed as

$$\begin{cases} \dot{x}_i = \kappa \sum_{j \in N_i} \left(m_j(t) - m_i(t)\right) \\ \dot{y}_i = v, \quad i \in I_n \end{cases} \tag{5.1}$$

where κ refers to a positive constant, and N_i represents the set of neighbors of agent i. In addition, v denotes the vertical speed of agents along the y-axis.

$$m_i(t) = \iint_{\Omega_i(t)} \rho(x, y)dxdy$$

represents the workload in the subregion $\Omega_i(t)$ at time t. The idea behind the dynamics (5.1) is that the partition bars move at the same speed v along the y-axis. Meanwhile, each agent communicates with its neighbors nearby to share the workload information of their respective subregions, so that their partition bars also move along the x-axis to equally partition the workload in the region Ω (For Agent n, we have $\dot{x}_n = g_b'(y_n)\dot{y}_n = g_b'(y_n)v$, since its partition bar moves along the boundary $x = g_b(y)$ at the speed v along the y-axis). Eventually, the trajectories of partition bars

Table 5.1 Distributed sweep coverage algorithm

Input: $x_i(0), i \in \mathcal{I}_n$ **Output:** $m_i(t), t \in [0, T_p]$
Set parameters v, σ and ϵ
Compute T_p and $e_i(0) = m_i(0), i \in \mathcal{I}_n$
while $(e_i(t) > 0)$
if $(t \leq T_p)$
Compute the workload $m_i(t)$ in $\Omega_i(t)$
Obtain the workload $m_j(t)$ in $\Omega_j(t)$, $j \in \mathcal{N}_i$
Partition the region Ω with the dynamics (A.1)
end if
Sweep the workload in $\Omega_i(t)$ with the rate σ
Compute residual workload $e_i(t) = m_i(t) - \sigma t$
end while

form the boundaries between adjacent subregions, and thus the region Ω is divided into n subregions (i.e., $\Omega_i, i \in \mathcal{I}_n$). To save the sweep time, each agent is required to complete the workload in their respective subregions at the constant sweeping rate σ, while partitioning the workload in the region with the dynamics (5.1).

Remark 5.1 At the beginning, all agents are located at the bottom of the region Ω (i.e., $y = 0$). When it starts the sweeping process, the partition bars of agents move towards the top of the region Ω (i.e., $y = l$) with the dynamics (5.1). In this way, the trajectories of partition bars succeed in dividing the coverage region Ω into subregions.

For simplicity, the distributed sweep coverage algorithm (DSCA) of multi-agent systems is summarized in Table 5.1. First of all, it is necessary to set the parameters v, σ and ϵ using constant values and compute the complete time of region partition $T_p = (l - \epsilon)/v$ and the initial error $e_i(0) = m_i(0), i \in \mathcal{I}_n$. Then each agent (say, agent i) starts to sweep its own subregion at the sweeping rate σ and meanwhile cooperates with its neighbors for the equal workload partition using workload memory. After the region partition is completed at time $t = T_p$, each agent focuses on sweeping the workload in its own subregion. Finally, the sweeping task is fulfilled after all the agents have cleaned up the workload in their respective subregions.

Remark 5.2 The DSCA is expected to sweep the workload in the general connected and bounded regions via coordinate transformation. Specifically, a homeomorphism $f : X \longrightarrow Y$ with the condition

$$\int_X \rho(x, y)dxdy = \int_Y \hat{\rho}(x, y)dxdy$$

can be constructed from the original bounded region X to the regular bounded region Y (e.g., a rectangular region). By implementing the DSCA in the regular region Y, the trajectories of partition bars (i.e., the boundaries between subregions) are mapped onto the original region X via the inverse function $f^{-1} : Y \longrightarrow X$, which allows to achieve the sweep coverage of the original region X.

5.3 Key Lemmas

To quantify the partition performance of multi-agent system with the dynamics (5.1), we introduce the mismatch vector at time t:

$$\Delta m(t) = m(t) - \bar{m}(t) \cdot \mathbf{1}_n, \tag{5.2}$$

where

$$m(t) = (m_1(t), m_2(t), ..., m_n(t))^T \in R^n$$

and $\mathbf{1}_n = (1, 1, ..., 1)^T \in R^n$. $\bar{m}(t)$ denotes the average workload on each subregion at time t, and it is computed as:

$$\bar{m}(t) = \frac{1}{n} \iint_{\Omega(t)} \rho(x, y) dx dy$$

with the partition region $\Omega(t) = \bigcup_{i=1}^{n} \Omega_i(t)$ at time t. Clearly, the workload is equally divided for each subregion when $\|\Delta m(t)\| = 0$ with the 2-norm $\| \cdot \|$. Next, three lemmas are provided to facilitate the derivation of main theoretical results in this paper. The logical path is presented as follows: Lemma 5.1 contributes to establishing the inequality in Lemma 5.3, which serves to demonstrate the input-to-sate stability of Multi-agent dynamics (5.1) in Theorem 5.1. Lemmas 5.2 and 5.3 are used to estimate the upper bound of time error ΔT in Theorem 5.2. In addition, Lemma 5.3 helps to derive the sufficient condition of collision avoidance among subregion boundaries in Theorem 5.3. The first lemma uncovers the relationship between the partition error of workload in subregions and the average workload in each subregion.

Lemma 5.1

$$\|\Delta m(t)\| \leq \sqrt{n(n-1)} \cdot \bar{m}(t).$$

Proof It follows from

$$\|\Delta m(t)\|^2 = \|m(t)\|^2 + n\bar{m}(t)^2 - 2\bar{m}(t)\sum_{i=1}^{n} m_i(t)$$
$$\leq \left(\sum_{i=1}^{n} m_i(t)\right)^2 - n\bar{m}(t)^2 = n(n-1)\bar{m}(t)^2$$

that

$$\|\Delta m(t)\| \leq \sqrt{n(n-1)} \cdot \bar{m}(t),$$

which completes the proof. □

Then we present the second lemma, which provides an estimation for the maximum partition error of workload in subregions, compared to the total partition error.

Lemma 5.2

$$\|\Delta m(t)\|_\infty \leq \sqrt{(n-1)/n} \cdot \|\Delta m(t)\|.$$

Proof First of all, we split $\|\Delta m(t)\|^2$ into two terms as follows

$$\|\Delta m(t)\|^2 = [\Delta m_k(t)]^2 + \sum_{i=1, i\neq k}^{n} [\Delta m_i(t)]^2, \quad \forall k \in \mathcal{I}_n. \tag{5.3}$$

For simplicity, define $z_i(t) = \Delta m_i(t)$, and we formulate the constrained optimization problem

$$\min \sum_{i=1, i\neq k}^{n} z_i(t)^2, \tag{5.4}$$

which is subject to $\sum_{i=1,i\neq k}^{n} z_i(t) = -z_k(t)$, since we have $\sum_{i=1}^{n} z_i(t) = 0$. To solve the optimization problem (5.4), we introduce the Lagrange function

$$\mathcal{L}(z_1, z_2, ..., z_n, c) = \sum_{i=1, i\neq k}^{n} z_i(t)^2 - c\sum_{i=1}^{n} z_i(t).$$

By solving the system of equations

$$\frac{\partial \mathcal{L}}{\partial c} = \sum_{i=1}^{n} z_i(t) = 0, \quad \frac{\partial \mathcal{L}}{\partial z_i} = 0, \quad i \in \mathcal{I}_n, \quad i \neq k,$$

we obtain $z_i(t) = -\frac{z_k(t)}{n-1}, i \in \mathcal{I}_n, i \neq k$ and the minimum of optimization problem (5.4)

$$\min \sum_{i=1, i\neq k}^{n+1} z_i(t)^2 = \frac{z_k(t)^2}{n-1},$$

which implies

$$\sum_{i=1, i\neq k}^{n} [\Delta m_i(t)]^2 \geq \frac{1}{n-1} [\Delta m_k(t)]^2 .$$

Then it follows from Eq. (5.3) that

$$\|\Delta m(t)\|^2 \geq [\Delta m_k(t)]^2 + \frac{1}{n-1} [\Delta m_k(t)]^2 = \frac{n}{n-1} [\Delta m_k(t)]^2 , \quad \forall k \in \mathcal{I}_n$$

which indicates

$$|\Delta m_k(t)| \leq \left(\frac{n-1}{n} \|\Delta m(t)\|^2 \right)^{\frac{1}{2}} , \quad \forall k \in \mathcal{I}_n .$$

Therefore, we conclude

$$\|\Delta m(t)\|_\infty = \max_{k\in\mathcal{I}_n} |\Delta m_k(t)| \leq \sqrt{(n-1)/n} \cdot \|\Delta m(t)\|.$$

The proof is thus completed. □

Remark 5.3 For $\Delta m(t) \in R^n$, there exists the relationship between the infinity norm and the Euclidean norm with the properties of norms as follows

$$\|\Delta m(t)\|_\infty \leq \|\Delta m(t)\|.$$

The relationship in Lemma 5.2 is less conservative according to

$$\|\Delta m(t)\|_\infty \leq \sqrt{(n-1)/n} \|\Delta m(t)\| \leq \|\Delta m(t)\|.$$

This is because the elements of $\Delta m(t)$ satisfy the constraint $\sum_{i=1}^{n} \Delta m_i(t) = 0$.

The partition error of workload in subregions evolves as the DSCA in Table 5.1 is implemented. The third lemma aims to estimate the partition error of workload with respect to time t, speed v, communication network and other parameters of the coverage region.

Lemma 5.3

$$\|\Delta m(t)\| \leq \|\Delta m(t_0)\| e^{-\frac{\kappa\lambda_2(t-t_0)}{2}} + \frac{1}{2} \int_{t_0}^{t} e^{\frac{\kappa\lambda_2(\tau-t)}{2}} \zeta(\tau) d\tau, \tag{5.5}$$

where λ_2 is the second smallest eigenvalue of a symmetric matrix and

$$\zeta(\tau) = 2v\bar{\rho} \left[\epsilon \sqrt{g_a'(y)^2 + g_b'(y)^2} + (g_b(y) - g_a(y)) \sqrt{(n-1)/n} \right] . \tag{5.6}$$

Proof The time derivative of $\|\Delta m(t)\|^2$ with respect to the trajectory of multi-agent dynamics (5.1) is given by

$$\frac{d\|\Delta m(t)\|^2}{dt} = 2\sum_{i=1}^{n} \Delta m_i(t) \left(\dot{x}_i \int_{y-\epsilon}^{y} \rho(x_i, y)dy - \dot{x}_{i-1} \int_{y-\epsilon}^{y} \rho(x_{i-1}, y)dy \right)$$
$$+ 2v \sum_{i=1}^{n} \Delta m_i(t) \left(\int_{x_{i-1}(y)}^{x_i(y)} \rho(x, y)dx - \frac{1}{n} \int_{g_a(y)}^{g_b(y)} \rho(x, y)dx \right). \tag{5.7}$$

The first term in Eq. (5.7) can be further expressed as

$$\sum_{i=1}^{n} \Delta m_i(t) \left(\dot{x}_i \int_{y-\epsilon}^{y} \rho(x_i, y)dy - \dot{x}_{i-1} \int_{y-\epsilon}^{y} \rho(x_{i-1}, y)dy \right)$$
$$= -\kappa \Delta m(t)^T G(x) L_{n-1} m(t) - v \Delta m(t)^T P(y)$$

where

$$P(y) = \left(g_a'(y) \int_{y-\epsilon}^{y} \rho(g_a(y), y)dy, 0, ..., 0, -g_b'(y) \int_{y-\epsilon}^{y} \rho(g_b(y), y)dy \right)^T \in R^n$$

and $L_{n-1} \in R^{(n-1)\times n}$ is composed of the first $(n-1)$ rows of Laplacian matrix that represents the communication network of multi-agent systems. In addition, $G(x) \in R^{n\times(n-1)}$ is given by

$$\begin{pmatrix} \int_{y-\epsilon}^{y} \rho(x_1, y)dy & 0 & . & . & . \\ -\int_{y-\epsilon}^{y} \rho(x_1, y)dy & \int_{y-\epsilon}^{y} \rho(x_2, y)dy & . & . & . \\ 0 & -\int_{y-\epsilon}^{y} \rho(x_2, y)dy & . & . & . \\ . & . & . & . & . \\ . & . & . & -\int_{y-\epsilon}^{y} \rho(x_{n-2}, y)dy & \int_{y-\epsilon}^{y} \rho(x_{n-1}, y)dy \\ . & . & . & 0 & -\int_{y-\epsilon}^{y} \rho(x_{n-1}, y)dy \end{pmatrix}$$

and it can be rewritten as

$$G(x) = -J_{-1,n}^T \cdot E_{n-1} \cdot \Lambda(\epsilon, x),$$

where $J_{-1,n}$ denotes the n-dimensional Jordan matrix with the diagonal elements being -1, and E_{n-1} is composed of the first $(n-1)$ columns in the n-dimensional unit matrix. Note that $\Lambda(\epsilon, x)$ refers to the $(n-1)$ dimensional diagonal matrix with the ith diagonal element being

$$\int_{y-\epsilon}^{y} \rho(x_i, y)dy, \quad i \in \mathcal{I}_{n-1}.$$

Because of

$$\Delta m(t)^T G(x) L_{n-1} m(t) = \Delta m(t)^T G(x) L_{n-1} \Delta m(t),$$

the first term in Eq. (5.7) can be estimated as

$$\begin{aligned}
&\sum_{i=1}^{n} \Delta m_i(t) \left(\dot{x}_i \int_{y-\epsilon}^{y} \rho(x_i, y) dy - \dot{x}_{i-1} \int_{y-\epsilon}^{y} \rho(x_{i-1}, y) dy \right) \\
&\leq -\kappa \Delta m(t)^T G(x) L_{n-1} \Delta m(t) - v \Delta m(t)^T P(y) \\
&= -\kappa \lambda_2 \|\Delta m(t)\|^2 - v \Delta m(t)^T P(y) \\
&\leq -\kappa \lambda_2 \|\Delta m(t)\|^2 + v \|\Delta m(t)\| \cdot \|P(y)\| \\
&\leq -\kappa \lambda_2 \|\Delta m(t)\|^2 + v \epsilon \bar{\rho} \sqrt{g_a'(y)^2 + g_b'(y)^2} \cdot \|\Delta m(t)\|
\end{aligned}$$

where λ_2 denotes the second smallest eigenvalue of the matrix

$$\frac{G(x) L_{n-1} + L_{n-1}^T G(x)^T}{2}$$

with $x = (x_1, x_2, ..., x_{n-1})$ and

$$x_i \in [\inf_{0 \leq y \leq l} g_a(y), \sup_{0 \leq y \leq l} g_b(y)], \quad i \in \mathcal{I}_{n-1}.$$

For the second term in Eq. (5.7), it follows from Cauchy–Schwarz inequality and Lemma 5.1 that

$$\begin{aligned}
&\sum_{i=1}^{n} \Delta m_i(t) \left(\int_{x_{i-1}(y)}^{x_i(y)} \rho(x, y) dx - \frac{1}{n} \int_{g_a(y)}^{g_b(y)} \rho(x, y) dx \right) \\
&\leq \|\Delta m(t)\| \left[\sum_{i=1}^{n} \left(\int_{x_{i-1}(y)}^{x_i(y)} \rho(x, y) dx - \frac{1}{n} \int_{g_a(y)}^{g_b(y)} \rho(x, y) dx \right)^2 \right]^{\frac{1}{2}} \\
&\leq \|\Delta m(t)\| \cdot \sqrt{(n-1)/n} \cdot \bar{\rho} \, [g_b(y) - g_a(y)] .
\end{aligned}$$

Therefore, the time derivative of $\|\Delta m(t)\|^2$ can be estimated as follows

$$\begin{aligned}
\frac{d \|\Delta m(t)\|^2}{dt} &\leq -\kappa \lambda_2 \|\Delta m(t)\|^2 + 2 v \bar{\rho} \epsilon \sqrt{g_a'(y)^2 + g_b'(y)^2} \cdot \|\Delta m(t)\| \\
&\quad + 2 v \bar{\rho} \sqrt{(n-1)/n} \cdot [g_b(y) - g_a(y)] \cdot \|\Delta m(t)\| \\
&= -\kappa \lambda_2 \|\Delta m(t)\|^2 + \zeta(t) \|\Delta m(t)\| .
\end{aligned} \tag{5.8}$$

with

$$\zeta(t) = 2v\bar{\rho}\left[\epsilon\sqrt{g_a'(y)^2 + g_b'(y)^2} + (g_b(y) - g_a(y))\sqrt{(n-1)/n}\right].$$

Solving the above differential inequality yields

$$\|\Delta m(t)\| \le \|\Delta m(t_0)\| e^{-\frac{\kappa\lambda_2(t-t_0)}{2}} + \frac{1}{2}\int_{t_0}^{t} e^{\frac{\kappa\lambda_2(\tau-t)}{2}}\zeta(\tau)d\tau.$$

The proof is thus completed. □

Remark 5.4 λ_2 is the second smallest eigenvalue of a symmetric matrix that depends on both communication network of multi-agent systems and the workload distribution on the coverage region. Mathematically, it can be expressed as

$$\lambda_2 = \inf_{x_i \in C} \lambda_2\left(\frac{G(x)L_{n-1} + L_{n-1}^T G(x)^T}{2}\right)$$

with

$$x = (x_1, x_2, ..., x_{n-1}), \quad C = [\inf_{0\le y\le l} g_a(y), \sup_{0\le y\le l} g_b(y)]$$

and

$$G(x) = -J_{-1,n}^T \cdot E_{n-1} \cdot \Lambda(\epsilon, x).$$

Here, $J_{-1,n}$ denotes the n-dimensional Jordan matrix with the diagonal elements being -1, and E_{n-1} is composed of the first $(n-1)$ columns in the n-dimensional unit matrix. In addition, $\Lambda(\epsilon, x)$ refers to the $(n-1)$ dimensional diagonal matrix with the ith diagonal element being

$$\int_{y-\epsilon}^{y} \rho(x_i, y)dy, \quad i \in \mathcal{I}_{n-1}.$$

Finally, $L_{n-1} \in R^{(n-1)\times n}$ is composed of the first $(n-1)$ rows of Laplacian matrix that represents the communication network of multi-agent systems. To guarantee the partition performance, the topology structure of community network has to be predetermined such that $\lambda_2 > 0$ holds.

5.4 Stability Analysis

This section provides main theoretical results for the DSCA in Table 5.1 with the aid of lemmas in Sect. 5.3. First of all, it is proved that multi-agent system (A.1) is input-to-state stable. Then the upper bound of ΔT is estimated for completing the workload in the coverage region Ω, followed by a sufficient condition to guarantee the collision avoidance of subregion boundaries. Finally, some discussions are presented in the remarks.

By treating the speed v of partition bars along the y-axis as the external input, we demonstrate the input-to-state stability of multi-agent system (5.1) as follows.

Theorem 5.1 *Multi-agent system (5.1) is input-to-state stable.*

Proof By treating $\Delta m(t)$ as the state vector of multi-agent system (5.1), Inequality (5.5) in Lemma 5.3 can be rewritten as

$$\begin{aligned}\|\Delta m(t)\| &\leq \|\Delta m(t_0)\| e^{-\frac{\kappa\lambda_2(t-t_0)}{2}} + \frac{1}{2}\int_{t_0}^{t} e^{\frac{\kappa\lambda_2(\tau-t)}{2}} \zeta(\tau)d\tau \\ &\leq \|\Delta m(t_0)\| e^{-\frac{\kappa\lambda_2(t-t_0)}{2}} + \frac{\bar{\zeta}}{\kappa\lambda_2},\end{aligned}$$

where

$$\bar{\zeta} = 2\bar{\rho}\|v\|_{[t_0,t]} \sup_{y\in[0,l]} \left[\epsilon\sqrt{g_a'(y)^2 + g_b'(y)^2} + (g_b(y) - g_a(y))\sqrt{(n-1)/n}\right]$$

and

$$\|v\|_{[t_0,t]} = \sup_{\tau\in[t_0,t]} \|v(\tau)\|.$$

Clearly, $\bar{\zeta}$ is a continuous increasing function with respect to the speed v, and it satisfies $\bar{\zeta} = 0$ when $v = 0$. Therefore, it follows from the definition in [18] that multi-agent system (5.1) is input-to-state stable when the speed v is regarded as the external input. □

For the DSCA described in Table 5.1, we provide a quantitative estimation for the upper bound of time error ΔT, which allows us to assess the sweep coverage performance of multi-agent systems.

Theorem 5.2 *With the DSCA in Table 5.1, the time error ΔT for sweeping the coverage region Ω is bounded by*

$$\Delta T \leq \frac{1}{\sigma}\sqrt{(n-1)/n}\left(\|\Delta m(0)\| e^{-\frac{\kappa\lambda_2 T_p}{2}} + \frac{1}{2}\int_0^{T_p} e^{\frac{\kappa\lambda_2(\tau-T_p)}{2}} \zeta(\tau)d\tau\right). \tag{5.9}$$

where $\zeta(\tau)$ is given by Eq. (5.6).

Proof Since the sweeping rate σ is the same for all agents, ΔT depends on the subregion with the maximum workload in subregions. Thus, it can be computed by

$$\Delta T = \frac{1}{\sigma}\left(\max_{i\in I_n} m_i(T_p) - \bar{m}(T_p)\right) \leq \frac{1}{\sigma}\max_{i\in I_n} |m_i(T_p) - \bar{m}(T_p)| = \frac{1}{\sigma}\|\Delta m(T_p)\|_\infty.$$

From Lemmas 5.2 and 5.3 with $t_0 = 0$, we have

$$\begin{aligned}\Delta T &\leq \frac{1}{\sigma}\sqrt{(n-1)/n} \cdot \|\Delta m(T_p)\| \\ &\leq \frac{1}{\sigma}\sqrt{(n-1)/n}\left(\|\Delta m(0)\|e^{-\frac{\kappa\lambda_2 T_p}{2}} + \frac{1}{2}\int_0^{T_p} e^{\frac{\kappa\lambda_2(\tau - T_p)}{2}}\zeta(\tau)d\tau\right).\end{aligned}$$

The proof is thus completed. □

Remark 5.5 Compared with existing results in [9], the proposed sweep coverage algorithm in this work can be implemented in a distributed manner to complete the workload without the centralized control command. When the vertical speed v is relatively low and ϵ is less than the stripe width d, Theorem 5.2 in this work can provide a tighter estimation on the upper bound of time error ΔT, compared to Theorem 4.1 in [9].

Remark 5.6 A tighter upper bound for time error ΔT can be obtained when ϵ is sufficiently small. This is due to $\lim_{\epsilon \to 0} \|\Delta m(0)\| = 0$, which leads to a tighter upper bound as follows

$$\Delta T \leq \frac{\bar{\rho}v(n-1)}{\sigma n}\int_0^{T_p} e^{\frac{\kappa\lambda_2(\tau - T_p)}{2}}\left[g_b(v\tau) - g_a(v\tau)\right]d\tau,$$

according to Eq. (5.9) in Theorem 5.2. Nevertheless, if the parameter ϵ is equal to 0, the partition bar degrades into a point and fails to effectively balance the workload in subregions during the sweeping process. Thus, ϵ should be positive and sufficiently small for a tight upper bound of time error ΔT.

Remark 5.7 To facilitate the estimation of time error ΔT, the time interval $[0, T_p]$ can be divided into a series of consecutive subintervals

$$c_i = [(i-1)T_p/q,\ iT_p/q], \quad 1 \leq i \leq q$$

and $q \in Z^+$. Considering that

$$e^{\frac{\kappa\lambda_2(\tau - T_p)}{2}}, \quad 0 \leq \tau \leq T_p$$

is an increasing function with respect to τ, we have

$$\begin{aligned}\int_0^{T_p} e^{\frac{\kappa\lambda_2(\tau - T_p)}{2n}}\zeta(\tau)d\tau &\leq \sum_{i=1}^{q} e^{\frac{\kappa\lambda_2(t_i - T_p)}{2}}\int_{t_{i-1}}^{t_i}\zeta(\tau)d\tau \\ &= \sum_{i=1}^{q} e^{-\frac{\kappa(q-i)\lambda_2 T_p}{2q}}\int_{(i-1)T_p/q}^{iT_p/q}\zeta(\tau)d\tau.\end{aligned}$$

which leads to

$$\Delta T \leq \frac{1}{\sigma}\sqrt{(n-1)/n}\left(\|\Delta m(0)\|e^{-\frac{\kappa\lambda_2 T_p}{2}} + \frac{T_p}{2q}\sum_{i=1}^{q} e^{-\frac{\kappa(q-i)\lambda_2 T_p}{2q}} \sup_{\tau\in c_i}\zeta(\tau)\right). \quad (5.10)$$

The upper bound of time error ΔT in Inequality (5.10) converges to that in Inequality (5.9) as the parameter q goes to the positive infinity. This implies that a tighter upper bound of time error ΔT can be achieved when the parameter q is relatively large. In practice, the parameter q can be increased gradually in order to obtain the desired estimation for the upper bound of ΔT.

While partitioning the coverage region Ω, the partition bars are not allowed to intersect with each other for achieving the disjoint subregions. Thus a sufficient condition is provided to avoid the collision of partition bars.

Theorem 5.3 *Collision avoidance of partition bars is guaranteed if the following inequality*

$$x_{i+1}(0) > x_i(0) + \kappa\|(e_{i+1}-e_i)^T L\| T_p\left(\|\Delta m(0)\| + \frac{1}{\kappa\lambda_2}\sup_{0\leq s\leq T_p}\zeta(s)\right), \quad \forall i \in \mathcal{I}_{n-2}$$

holds, where $\zeta(\tau)$ is given by Eq. (5.6) and e_i denotes the ith column vector in the n dimensional unit matrix. In addition, L represents the Laplacian matrix of the multi-agent communication network.

Proof Let $\Delta x_i(t) = x_{i+1}(t) - x_i(t),\ i \in \mathcal{I}_{n-2}$ denote the distance between two adjacent partition bars at time t. Then we have

$$\frac{d\Delta x_i(t)}{dt} = \dot{x}_{i+1}(t) - \dot{x}_i(t) = \kappa e_i^T Lm(t) - \kappa e_{i+1}^T Lm(t) = \kappa(e_i - e_{i+1})^T Lm(t)$$

It follows from $Lm(t) = L\Delta m(t)$ that

$$\frac{d\Delta x_i(t)}{dt} = -\kappa(e_{i+1}-e_i)^T L\Delta m(t).$$

Therefore, we have

$$\begin{aligned}\Delta x_i(t) &= \Delta x_i(0) - \kappa\int_0^t (e_{i+1}-e_i)^T L\Delta m(\tau)d\tau \\ &= \Delta x_i(0) - \kappa(e_{i+1}-e_i)^T L\int_0^t \Delta m(\tau)d\tau \\ &\geq \Delta x_i(0) - \kappa\|(e_{i+1}-e_i)^T L\|\cdot\left\|\int_0^t \Delta m(\tau)d\tau\right\|\end{aligned}$$

which leads to

$$
\begin{aligned}
&\Delta x_i(t)\\
&\geq \Delta x_i(0) - \kappa \|(e_{i+1} - e_i)^T L\| \cdot \int_0^t \|\Delta m(\tau)\| d\tau\\
&\geq \Delta x_i(0) - \kappa \|(e_{i+1} - e_i)^T L\| \cdot \int_0^t \left(\|\Delta m(0)\| e^{-\frac{\kappa\lambda_2 \tau}{2}} + \frac{1}{2}\int_0^\tau e^{\frac{\kappa\lambda_2(s-\tau)}{2}} \zeta(s) ds \right) d\tau\\
&\geq \Delta x_i(0) - \kappa \|(e_{i+1} - e_i)^T L\| \cdot \int_0^t \left(\|\Delta m(0)\| + \frac{1}{\kappa\lambda_2} \sup_{0\leq s\leq \tau} \zeta(s) \right) d\tau\\
&\geq \Delta x_i(0) - \kappa \|(e_{i+1} - e_i)^T L\| T_p \left(\|\Delta m(0)\| + \frac{1}{\kappa\lambda_2} \sup_{0\leq s\leq T_p} \zeta(s) \right)
\end{aligned}
$$

according to

$$
\left\| \int_0^t \Delta m(\tau) d\tau \right\| \leq \int_0^t \|\Delta m(\tau)\| d\tau
$$

and Lemma 5.3. To avoid the collision of partition bars, we need to ensure

$$
\Delta x_i(t) > 0, \forall t \in [0, T_p], \; i \in \mathcal{I}_{n-2},
$$

which can be achieved by the following inequality

$$
\Delta x_i(0) - \kappa \|(e_{i+1} - e_i)^T L\| T_p \left(\|\Delta m(0)\| + \frac{1}{\kappa\lambda_2} \sup_{0\leq s\leq T_p} \zeta(s) \right) > 0, \quad \forall i \in \mathcal{I}_{n-2}.
$$

The proof is thus completed. □

Remark 5.8 During the partition, the outmost two partition bars (i.e., x_1 and x_{n-1}) have to avoid the collision against the boundaries $g_a(y)$ and $g_b(y)$, respectively. Then the sufficient conditions for avoiding the boundary collision are given by

$$
x_1(0) > \sup_{0\leq y\leq l} g_a(y) + \kappa \|e_1^T L\| T_p \left(\|\Delta m(0)\| + \frac{1}{\kappa\lambda_2} \sup_{0\leq s\leq T_p} \zeta(s) \right)
$$

and

$$
\inf_{0\leq y\leq l} g_b(y) > x_{n-1}(0) + \kappa \|e_{n-1}^T L\| T_p \left(\|\Delta m(0)\| + \frac{1}{\kappa\lambda_2} \sup_{0\leq s\leq T_p} \zeta(s) \right),
$$

respectively.

Remark 5.9 For the rectangular region of length l_α and width l_β, when the length of partition bars ϵ goes to zero, the upper bound of ΔT can be estimated by

$$\Delta T \le \frac{2l_\beta \bar{\rho} v}{\kappa \sigma \lambda_2} \cdot \frac{n-1}{n} \cdot \left(1 - e^{-\kappa \lambda_2 T_p/2}\right) \le \frac{2l_\beta \bar{\rho} v}{\kappa \sigma \lambda_2} \cdot \frac{n-1}{n}, \quad \epsilon \to 0$$

Furthermore, if the speed v approaches zero as well, we have

$$0 \le \Delta T \le \frac{2l_\beta \bar{\rho} v}{\kappa \sigma \lambda_2} \cdot \frac{n-1}{n} \to 0, \quad v \to 0$$

This implies that the actual sweeping time T approaches the optimal sweeping time T^* as both the speed v and the length of partition bars ϵ go to zero.

Remark 5.10 During the sweeping process, it is assumed that the inequality $\sigma t \le m_i(t)$ holds for $t \in [0, T_p]$, $i \in \mathcal{I}_n$, which implies that the sweeping operation lags that of region partition, and the agents do not complete the workload in their respective subregion before the region partition comes to an end.

Remark 5.11 It is possible to extend the proposed sweep coverage approach to the surface in the three dimensional Euclidean space. The dynamics of partition bars in the first two dimensions (i.e., xy-plane) can be the same as Eq. (5.1), and it determines the dynamics of partition bars in the third dimension (i.e., z-axis) due to the surface constraint. In addition, we can estimate the time error between the actual sweep time and the optimal time for sweeping the surface in the same way.

5.5 Numerical Simulations

This section provides numerical examples to validates the DSCA in Table 5.1 and theoretical results on the time error ΔT. Specifically, 5 mobile agents are connected in a directed chain-like communication network (from the 5th agent to the 1st agent) to cooperatively sweep the region Ω, which is enclosed by two curves:

$$x_0 = g_a(y) = 0.2 \sin \frac{\pi(y-4)}{3} + 1, \quad x_5 = g_b(y) = 0.2 \sin \frac{\pi(y-4)}{3} + 6$$

and two straight lines: $y = 0$ and $y = 10$. In addition, the distribution density function of workload is described by

$$\rho(x, y) = \frac{3}{2} + \frac{1}{2} \sin \frac{x^2 + y^2}{5}$$

with the upper bound $\bar{\rho} = 2$ and the lower bound $\underline{\rho} = 1$. Other parameters are given as follows: $\kappa = 1$, $\epsilon = 0.01$, $v = 8$, $\sigma = 6$, $q = 10$ and $\lambda_2 = 0.0011$. For simplicity, the Euler method is employed to implement the partition dynamics (5.1) with a step size of 0.001. Figure 5.2 presents the cooperative sweep process of 5 mobile agents in the coverage region, where the color bar indicates the workload density ranging from light yellow to dark red. At the initial time $t = 0$, all the partition bars

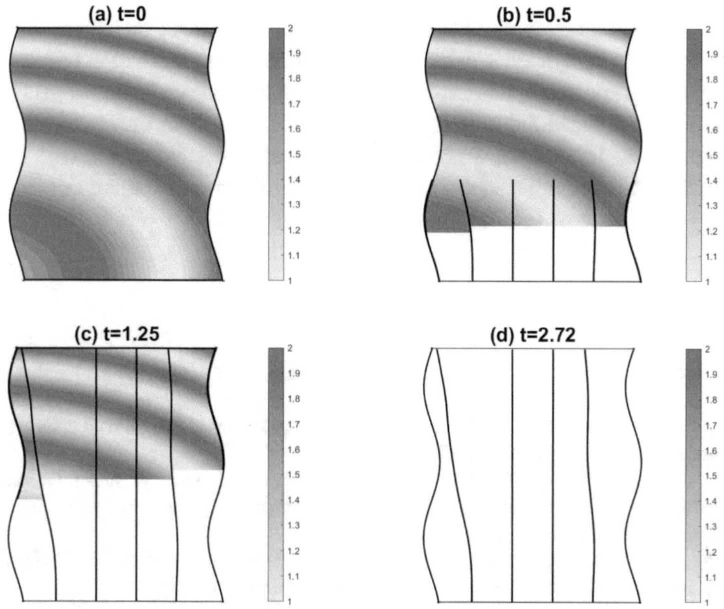

Fig. 5.2 Numerical example on distributed sweep coverage of 5 mobile agents

are located at the bottom of the region. Then multi-agent system starts partitioning the whole region and meanwhile the agents are sweeping their own subregions. At $t = 0.5$, part of the region has been partitioned by the trajectories of partition pars, and each agent completes the same workload $\sigma t = 3$ in its own subregion. The completed parts are marked in white. The task of region partition is finished at $t = 1.25$ when all the partition bars arrive at the top of coverage region. Next, the mobile agents continue to sweep their respective subregions. Finally, the sweeping process comes to an end at $t = 2.72$ after cleaning up the workload in the whole region. Since the optimal sweeping time T^* is equal to 2.54, the time error is $\Delta T = 0.18$. According to Theorem 5.2 in this work, the upper bound of time error ΔT is 0.88. In comparison, the upper bound of time error ΔT is 13.29 using Theorem 4.1 in [9]. This demonstrates that the theoretical results in this work can provide a tighter upper bound of time error ΔT in comparison with existing results [9].

5.6 Conclusions

In this chapter, we developed a novel formulation for the sweep coverage of multi-agent systems in the uncertain environment. It has been proved that the dynamics of multi-agent systems is input-to-state stable. Moreover, we obtained the upper bound for the error between the actual sweep time and the optimal sweep time. A sufficient condition was derived to avoid the collision of partition bars in the process of workload partition. Numerical simulations demonstrated the effectiveness of the proposed sweep approach. It is expected that the proposed sweep coverage algorithm can complete the workload in the general region via the coordinate transformation. Future work may include the sweep coverage of an uncertain region with obstacles and the kinematics of multi-agent system with nonholonomic constraints.

References

1. Ni, W., Cheng, D.: Leader-following consensus of multi-agent systems under fixed and switching topologies. Syst. Control Lett. **59**(3–4), 209–217 (2010)
2. Hong, Y., Hu, J., Gao, L.: Tracking control for multi-agent consensus with an active leader and variable topology. Automatica **42**(7), 1177–1182 (2006)
3. Wen, G., Hu, G., Yu, W., Cao, J., Chen, G.: Consensus tracking for higher-order multi-agent systems with switching directed topologies and occasionally missing control inputs. Syst. Control Lett. **62**(12), 1151–1158 (2013)
4. Zhang, H.T., Zhai, C., Chen, Z.: A general alignment repulsion algorithm for flocking of multi-agent systems. IEEE Trans. Autom. Control **56**(2), 430–435 (2011)
5. Su, H., Wang, X., Lin, Z.: Flocking of multi-agents with a virtual leader. IEEE Trans. Autom. Control **54**(2), 293–307 (2009)
6. Alberton, R., Carli, R., Cenedese, A., Schenato, L.: Multi-agent perimeter patrolling subject to mobility constraints. In: Proceedings of American Control Conference, pp. 4498–4503 (2012)
7. Hussein, I.I., Stipanovic, D.M.: Effective coverage control for mobile sensor networks with guaranteed collision avoidance. IEEE Trans. Control Syst. Technol. **15**(4), 642–657 (2007)
8. Song, C., Liu, L., Feng, G., Xu, S.: Optimal control for multi-agent persistent monitoring. Automatica **50**(6), 1663–1668 (2014)
9. Zhai, C., Hong, Y.: Decentralized sweep coverage algorithm for multi-agent systems with workload uncertainties. Automatica **49**(7), 2154–2159 (2013)
10. Choset, H.: Coverage for robotics: a survey of recent results. Ann. Math. Artif. Intell. **31**(1), 113–126 (2001)
11. Cortés, J., Martínez, S., Karatas, T., Bullo, F.: Coverage control for mobile sensing network. IEEE Trans. Robot. Autom. **20**(2), 243–255 (2004)
12. Luna, J., Fierro, R., Abdallah, C., Wood, J.: An adaptive coverage control algorithm for deployment of nonholonomic mobile sensors. In: Proceedings of IEEE Conference on Decision and Control, pp. 1250–1256. Atlanta (2010)
13. Gage, D.W.: Command control for many-robot systems. In: Proceedings of Annual AUVS Teachnical Symposium, pp. 22–24 (1992)
14. Ghosh, A., Das, S.K.: Coverage and connectivity issues in wireless sensor networks: a survey. Pervasive Mob. Comput. **4**(3), 303–334 (2008)
15. Kumar, S., Lai, T.H., Arora, A.: Barrier coverage with wireless sensors. Wireless Netw. **13**(6), 817–834 (2007)
16. Cheng, T.M., Savkin, A.V.: Decentralized coordinated control of a vehicle network for deployment in sweep coverage. In: Proceedings of IEEE International Conference on Control and Automation, pp. 275–279 (2009)
17. Jeremic, A., Nehorai, A.: Design of chemical sensor arrays for monitoring disposal sites on the ocean floor. IEEE J. Oceanic Eng. **23**(4), 334–343 (1998)
18. Khalil, H.K.: Nonlinear Systems. Prentice-Hall, New Jersey (1996)

Chapter 6
Cooperative Sweep Coverage Algorithm of Discrete Time Multi-agent Systems

6.1 Introduction

Multi-agent systems and control have received increasingly attention in a wide variety of applications, which play an important role in distributed collection and control information. Distributed design with advantages such as low cost, reliability, and flexibility provides a feasible way to deploy a large number of networked agents over a region of interest to achieve desired collective tasks. In practice, agents are usually equipped with various sensors. Since a single agent may be difficult to complete the task due to its limited capacities, a group of agents (or viewed as a mobile sensor network) are usually teamed up to complete the tasks by communicating and coordinating their actions through network. Various coordination tasks and algorithms for multiple agents have been reported [1–4].

Cooperative coverage problems of multiple agents have drawn much attention to the researchers in recent years, particularly for robotic networks [5, 6]. There are various dynamic coverage types including barrier coverage, sweep coverage, and blanket coverage [7]. Different from the set containment to drive the agents to a give set [8], the coverage problem is to design multi-agent dynamics within a given region. Compared with the static coverage [9, 10], where the Voronoi partition was adopted with the locational cost function as an index to optimize sensor locations, dynamic coverage is considered to allow agents move around to cover the region of interest. Sweep coverage as a dynamic coverage problem is to make a group of agents with sensing capability move across the given region to detect targets of interest or complete workload on the area. It is difficult in that all the agents have to cooperate in order to optimize the operation time. Reference [11] proposed a formation-based method sweep coverage strategy. Additionally, [12] proposed a sweep coverage algorithm for the rectangular region based on the nearest neighbor's information. However, many problems of the sweep coverage with environmental uncertainties in any bounded region remains to be solved.

The objective of this chapter is to propose a discrete time sweep algorithm in the uncertain environment with different communication topologies. Although the

C. Zhai et al., *Cooperative Coverage Control of Multi-Agent Systems and its Applications*, Studies in Systems, Decision and Control 408,
https://doi.org/10.1007/978-981-16-7625-3_6

parameter uncertainty in the environment was studied based on adaptive control to learn environment information online in [6], we consider bounded uncertainties and cannot apply adaptive technique here. Because of those uncertainties, we cannot give a fixed formation-based coverage strategy in advance. Instead, we have to deal with the uncertainty when we carry out the coverage control. Thus, we propose a decentralized sweep coverage algorithm with two combined operations: partition (to handle the uncertainties) and sweep (to complete the coverage). Here a decentralized control technique is adopted to deal with the unknown dynamical environment. Although it is impossible to achieve sweep coverage of the given region with optimal operation time due to the uncertainty, we give the estimation on the difference between the actual coverage time and the optimal time.

The rest of the paper is organized as follows. A discrete time formulation of sweep coverage in uncertain environment is presented in Sect. 6.2. Then, a decentralized sweep coverage algorithm and the estimation of the extra time are shown in Sect. 6.3, followed by simulation results in Sect. 6.4. Finally, conclusions are given in the last section.

6.2 Problem Formulation

In this chapter, we consider the problem of sweeping the region with unknown workload distribution by a multi-agent system. The basic problem setup was given in [12], but we still introduce it to be self-contained. To increase the effectiveness in the coverage control, the whole region is divided into several subregions, and each agent is responsible for completing workload on its own subregion. If there is no uncertainty and workload distribution can be known for all the agents in advance, the optimal strategy to complete the coverage can be carried out by partitioning the whole region into subregions with equal workload for each agent, where the sweeping task can be completed in the shortest time regardless of the shape of subregions. Unfortunately, the workload distribution in an uncertain environment is not known in advance. Limited sensing range of each agent and unknown workload distribution make it impossible to get average workload on the whole region for each agent at each moment.

Consider a rectangular region D with width l_a and length l_b (see Fig. 6.1). All the agents line up at the bottom of D, and they will move to the top by sweeping all the rectangular region. Suppose each agent has its actuation range with diameter d, so when it sweeps when it moves, it will clean up a stripe with width d. For simplicity, the whole region is partitioned into stripes with length l_a and width d, and the agents sweep these stripes one by one to complete the whole region coverage. For convenience, we assume $l_b = qd$ for some integer q. To shorten the whole time for coverage, each stripe is partitioned into sub-stripes for each agent so that each agent has the same workload. Then each agent sweeps its own sub-stripe and simultaneously partitions the next stripes for sweep.

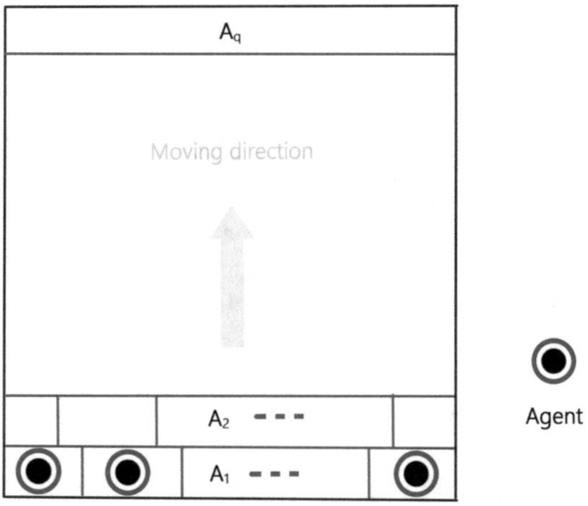

Fig. 6.1 Sweep coverage of multiple mobile agents on the rectangular region D

The workload distribution on D is denoted by $\rho(x, y)$, which is continuous and always larger than 0 on D. With respect to $\rho(x, y)$, we have the following assumption as in [12]. There exist positive constants $\underline{\rho}$ and $\bar{\rho}$ such that

$$\underline{\rho} \leq \rho(x, y) \leq \bar{\rho}.$$

For simplicity, we suppose the first stripe is partitioned into sub-stripes with equal workload for each agent. Moreover, each agent has the same sweeping rate v. Obviously, the completion time of sweeping each sub-stripe only depends on v and workload on the sub-stripe without considering the area and shape of sub-stripe. Moreover, each agent detects the workload around its position and gets the workload information of its neighbors on the next stripe. When each agent sweeps its own sub-stripe, the partition on the next stripe is conducted according to the proposed partition algorithm. Once all the agents finish sweeping their own sub-stripes, partition operation stops. Then, all the agents move to the newly partitioned stripes and repeat previous operations until the whole sweeping task is completed. Naturally, the workload on the ith sub-stripe of a stripe (say A_k) is given by

$$m_i^k = \int_{x_{i-1}^k}^{x_i^k} \omega_k(\tau) d\tau$$

where

$$\omega_k(\tau) = \int_{y_{k-1}}^{y_k} \rho(\tau, y) dy$$

with $y_0 = 0$, $y_q = l_b$ and $y_k = y_{k-1} + d$, $1 \leq k \leq q$. Clearly, x_{i-1}^k and x_i^k are horizontal positions of partition marks $i-1$ and i on A_k, respectively (where $i \in E_n = \{1, 2, ..., n\}$, $x_0^k = 0$ and $x_n^k = l_a$). The discrete time dynamic partition on A_k of

Table 6.1 Discrete time sweep coverage algorithm

For $i \in E_n$, i-th agent performs as follow.
set $k = 1$
while A_k is not the last stripe on D **do**
while workload on A_k is unfinished **do**
if m_i^k is not completed **then**
partition workload on A_{k+1} with (6.1)
sweep i-th sub-stripe on A_k
else
partition workload on A_{k+1} with (6.1)
end if
end while
stop partition and move to i-th sub-stripe on A_{k+1}
set $k = k + 1$
end while
sweep i-th sub-stripe on A_k until the workload is finished

partition mark i is given as follows

$$\frac{x_i^k(zT_s + T_s) - x_i^k(zT_s)}{T_s} = u_i(zT_s), \quad i = 1, ..., n-1 \tag{6.1}$$

where T_s denotes the sampling period, $z \in N^+$, and $u_i(zT_s)$ is the partition control input as follows

$$u_i(zT_s) = \kappa \sum_{j \in \mathcal{N}_i} (m_j^k(zT_s) - m_i^k(zT_s)), \tag{6.2}$$

Here, $\kappa > 0$ is a given constant and $\mathcal{N}_i$ represents the neighbor set of agent i. For simplicity of notation, T_s in brackets is omitted when no confusion is caused. Table 6.1 describes the discrete time sweep coverage algorithm for ith agent, which is similar to the distributed sweep coverage algorithm in [12] except for the discrete time partition dynamics.

6.3 Main Results

In this section, we give the theoretical analysis for the proposed sweep coverage algorithm. The basic idea is to estimate the extra time for the discrete time sweep coverage algorithm with the help of the continuous time model and some numerical methods since it is difficult to investigate the discrete time model directly (see Fig. 6.2). First of all, the workload partition algorithm for agent i on each stripe is presented as fol-

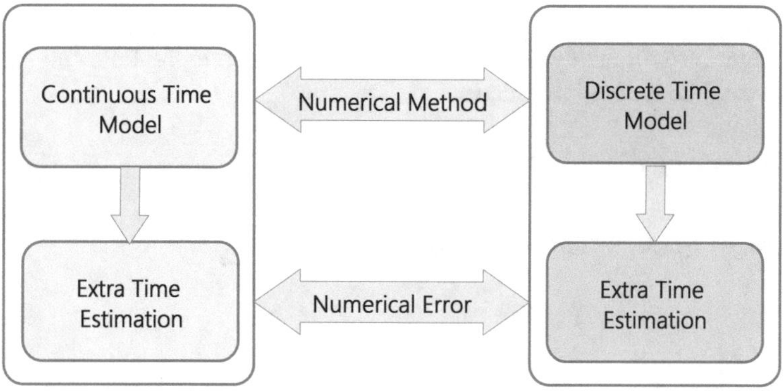

Fig. 6.2 Relationship between continuous time model and discrete time counterpart. The discrete time model can be obtained by discretizing the continuous time model with numerical methods such as Euler method. Moreover, the error caused by the discretization can be easily estimated. Then we can get the upper bound of the extra time for the discrete time model by using the error and results in continuous time model

lows: Each agent collects the workload information on neighbors' sub-stripes, and all the agents update the position of partition mark i with (6.1) and (6.2). Combining (6.1) and (6.2), we get the following dynamics of partition marks:

$$\frac{x_i^k(z+1) - x_i^k(z)}{T_s} = \kappa \sum_{j \in N_i} (m_j^k(z) - m_i^k(z)), \quad i = 1, ..., n-1 \tag{6.3}$$

Clearly, as the sampling period T_s tends to 0, we get the continuous time partition dynamics of mark i

$$\dot{x}_i^k = \kappa \sum_{j \in N_i} (m_j^k - m_i^k), \quad i = 1, ..., n-1 \tag{6.4}$$

Notice that the time derivative of m_i^k along (6.4) can be expressed as

$$\dot{m}_i^k = \omega_k(x_i^k)\dot{x}_i^k - \omega_k(x_{i-1}^k)\dot{x}_{i-1}^k, \quad i \in E_n \tag{6.5}$$

Rewrite the above set of equations in matrix form as follows.

$$\dot{m}^k = \begin{pmatrix} \sigma_1 & 0 & .. & 0 & 0 \\ -\sigma_1 & \sigma_2 & .. & 0 & 0 \\ 0 & -\sigma_2 & .. & 0 & 0 \\ . & . & .. & . & . \\ . & . & .. & . & . \\ 0 & 0 & .. & -\sigma_{n-2} & \sigma_{n-1} \\ 0 & 0 & .. & 0 & -\sigma_{n-1} \end{pmatrix} \dot{x}^k$$

where $x^k = (x_1^k, x_2^k, ..., x_{n-1}^k)^T$, $m^k = (m_1^k, m_2^k, ..., m_n^k)^T$ and $\sigma_i = \omega_k(x_i^k) > 0$. According to (6.4), the dynamics (6.5) can be converted to the following "linear" system with the state m^k as follows.

$$\begin{aligned}
\dot{m}_1^k &= -\kappa s_1 \sigma_1 m_1^k + \kappa \sigma_1 \sum_{j \in \mathcal{N}_1} m_j^k \\
\dot{m}_2^k &= -\kappa s_2 \sigma_2 m_2^k + \kappa s_1 \sigma_1 m_1^k - \kappa \sigma_1 \sum_{j \in \mathcal{N}_1} m_j^k + \kappa \sigma_2 \sum_{j \in \mathcal{N}_2} m_j^k \\
&\vdots \\
\dot{m}_{n-1}^k &= -\kappa s_{n-1} \sigma_{n-1} m_{n-1}^k + \kappa s_{n-2} \sigma_{n-2} m_{n-2}^k - \kappa \sigma_{n-2} \sum_{j \in \mathcal{N}_{n-2}} m_j^k + \kappa \sigma_{n-1} \sum_{j \in \mathcal{N}_{n-1}} m_j^k \\
\dot{m}_n^k &= \kappa s_{n-1} \sigma_{n-1} m_{n-1}^k - \kappa \sigma_{n-1} \sum_{j \in \mathcal{N}_{n-1}} m_j^k
\end{aligned} \tag{6.6}$$

where s_i denotes the number of agents in the neighbor set of agent i. Similarly, we rewrite the above equations in a compact form

$$\dot{m}^k = -\kappa P m^k, \tag{6.7}$$

where P is time-varying matrix based on x^k. Therefore, we will focus on the consensus problem of the time-varying system (6.7).

Clearly, the dynamical equation (6.7) can be viewed as a linear time-varying system since the elements of P depending on linear combination of σ_i that is related to x_i^k are time-varying. As for P, we have the following result.

Lemma 6.1

$$P\mathbf{1} = P^T\mathbf{1} = \mathbf{0}$$

where $\mathbf{1} = (1, 1, ..., 1)^T \in R^n$ *and* $\mathbf{0}$ *denotes the zero vector with n elements.*

Proof Take $m^k = \mathbf{1}$ and it follows from (6.6) and (6.7) that $-\kappa P\mathbf{1} = \mathbf{0}$. Since $\kappa > 0$, $P\mathbf{1} = \mathbf{0}$. Note that $P^T\mathbf{1} = \mathbf{0}$ means the sum of elements in each column of P is 0. Without loss of generality, we consider the sum of elements in the jth column of $P = [p_{ij}]$. Let $\mathcal{N}_\mathbf{j}$ represent the set of agents, which have the common neighbor, agent j. According to (6.6), we have

$$\sum_{i=1}^{n} p_{ij} = -s_j\sigma_j + s_j\sigma_j + \sum_{i \in \mathcal{N}_\mathbf{j}} (\sigma_i - \sigma_i) = 0$$

for $1 \leq j < n$, and

$$\sum_{i=1}^{n} p_{ij} = \sum_{i \in \mathcal{N}_\mathbf{j}} (\sigma_i - \sigma_i) = 0$$

for $j = n$. Thus, the proof is completed. □

Denote $P_u = \frac{P+P^T}{2}$, and 0 is the eigenvalue of the symmetric matrix P_u with the associated eigenvector $\mathbf{1}$, since

$$P_u\mathbf{1} = \frac{1}{2}(P\mathbf{1} + P^T\mathbf{1}) = \mathbf{0}.$$

Then we label n eigenvalues of P_u according to increasing order:

$$0 = \lambda_1(P_u) \leq \lambda_2(P_u) \leq \ldots \leq \lambda_{n-1}(P_u) \leq \lambda_n(P_u).$$

Here we assume that $\lambda_2(P_u) > 0$ holds in the following analysis.

Remark 6.1 The assumption $\lambda_2(P_u) > 0$ is not so restrictive, which is related to the connectivity (depending on the workload). If we select $u_i = \kappa(m_{i+1}^k - m_i^k)$ as the control input, the assumption holds immediately. Actually, the interconnection graph of agents in [12] is a directed chain, and the corresponding matrix P_u can be expressed as

$$\begin{pmatrix} \sigma_1 & -\sigma_1 & .. & 0 & 0 \\ -\sigma_1 & \sigma_1 + \sigma_2 & .. & 0 & 0 \\ . & . & .. & . & . \\ . & . & .. & . & . \\ 0 & 0 & .. & \sigma_{n-2} + \sigma_{n-1} & -\sigma_{n-1} \\ 0 & 0 & .. & -\sigma_{n-1} & \sigma_{n-1} \end{pmatrix}.$$

Clearly, P_u has only one zero eigenvalue with the eigenvector $\mathbf{1}$, since rank(P_u)= $n - 1$. Furthermore, all other eigenvalues of P_u expect for $\lambda_1(P_u) = 0$ are larger than zero according to Geršgorin disk theorem [13]. The assumption on $\lambda_2(P_u) > 0$ is thus satisfied.

Define

$$\lambda_2 = \min_{x^k \in \Delta^{n-1}} \lambda_2(P_u), \quad \Delta = [0, l_a]$$

Next, we give the result related to workload partition in the uncertain region. The following lemma provides the convergence analysis of workload partition algorithm on the stripe A_k.

Lemma 6.2 *The plane*

$$\{m^k \in R^n \mid m_1^k = m_2^k = \ldots = m_n^k\}$$

of the dynamics (6.1) with control input (6.2) is exponentially stable.

Proof To show the stability, we consider the change of variables $\delta_i = m_i^k - \bar{m}^k, i \in E_n$ and the following lower bounded function

$$H_k = \frac{1}{2}\sum_{i=1}^{n}(m_i^k - \bar{m}^k)^2, \quad \bar{m}^k = \frac{1}{n}\int_0^{l_a} \omega_k(\tau)d\tau$$

Thus, we have

$$H_k = \frac{1}{2}\sum_{i=1}^{n}\delta_i^2 = \frac{1}{2} \parallel \delta \parallel^2$$

where $\delta = (\delta_1, \delta_2, ..., \delta_n)^T$ and $\parallel \cdot \parallel$ denotes the Euclidean norm. Since

$$\delta = m^k - \bar{m}^k \mathbf{1}$$

and

$$P\delta = Pm^k - \bar{m}^k P\mathbf{1} = Pm^k$$

according to Lemma 6.1. Moreover, we have

$$\dot{\delta} = \dot{m}^k = -\kappa Pm^k = -\kappa P\delta.$$

The time derivative of H_k along the trajectories of (6.7) is given by

$$\dot{H}_k = \delta^T\dot{\delta} = -\kappa\delta^T P\delta.$$

Due to

$$\delta^T P\delta = \delta^T P^T\delta = \frac{1}{2}\delta^T(P + P^T)\delta = \delta^T P_u\delta,$$

we get

$$\dot{H}_k = -\kappa\delta^T P_u\delta$$

Further, $\delta^T P_u\delta$ has the following property [13].

$$\min_{\delta\perp\mathbf{1},\delta\neq\mathbf{0}} \frac{\delta^T P_u\delta}{\parallel \delta \parallel^2} = \lambda_2(P_u) \geq \lambda_2$$

Hence,

$$\dot{H}_k = -\kappa\delta^T P_u\delta \leq -\kappa\lambda_2 \parallel \delta \parallel^2 \leq -2\kappa\lambda_2 H_k.$$

Solving the above differential equation yields

$$H_k(t) \leq H_k(t_0)e^{-2\kappa\lambda_2(t-t_0)}, \tag{6.8}$$

which implies the conclusion. □

H_k defined in Lemma 6.2 is used to measure the uniformity of workload partition on A_k. Obviously, the smaller H_k is, the more well-proportioned workload partition we will get. In addition, H_k^a and H_k^b denote the uniformity of workload partition at initial and final partition positions on A_k, respectively. Note that the boundaries between sub-stripes on A_k are exactly the final positions of the partition marks on the stripe. Clearly, the initial positions of partition marks on A_{k+1} is the final partition position, that is, the boundaries between sub-stripes on A_k, respectively. Recalling Lemma 3.3 in [12], we can easily get the following lemma, and the proof is thus omitted here.

Lemma 6.3 *H_{k+1}^a and H_k^b satisfy the following inequality*

$$H_{k+1}^a \leq \alpha^2 H_k^b + \beta \frac{d^2 l_a^2}{n}, \tag{6.9}$$

where $\alpha = \frac{\bar{\rho}}{\underline{\rho}}$ and $\beta = \frac{\bar{\rho}^4}{\underline{\rho}^2} - \underline{\rho}^2$.

Let $x^k(t_z)$ and $x^k(z)$ denote the position vector of partition marks on stripe k $(1 \leq k \leq q)$ at time t_z in the continuous time formulation and discrete computation, respectively. Then we have the following result.

Lemma 6.4 *The global truncation error $e_z^k = x^k(t_z) - x^k(z)$ on stripe k satisfies*

$$\|e_z^k\| \leq \frac{(n+2)\kappa\bar{\rho}dl_a}{8}(\gamma - 1)T_s + \gamma\|e_0^k\|$$

where

$$\gamma = e^{\frac{4(n-1)\kappa d^2 \bar{\rho}^2 l_a}{\nu}}.$$

Proof In numerical computation, we adopt Euler method to implement the continuous time partition algorithm, which can be rewritten as follows

$$\begin{aligned}\dot{x}_i^k &= \kappa \sum_{j \in N_i} (m_j^k - m_i^k) = \kappa \sum_{j \in N_i} \left(\int_{x_{j-1}^k}^{x_j^k} \omega_k(\tau)d\tau - \int_{x_{i-1}^k}^{x_i^k} \omega_k(\tau)d\tau\right) \\ &= f_i(t, x^k), \quad i = 1, 2, ..., n-1\end{aligned}$$

with $x^k = (x_1^k, x_2^k, ..., x_n^k)^T$. From the mean value theorem, we have

$$f_i(t, x^k) - f_i(t, x^{k^*}) = \sum_{j=1}^{n} \frac{\partial f_i}{\partial x_j^k}(\xi_j)(x_j^k - x_j^{k^*})$$

Therefore, we get

$$
\begin{aligned}
|f_i(t, x^k) - f_i(t, x^{k^*})| &= |\sum_{j=1}^{n} \frac{\partial f_i}{\partial x_j^k}(\xi_j)(x_j^k - x_j^{k^*})| \\
&\le \sum_{j=1}^{n} |\frac{\partial f_i}{\partial x_j^k}(\xi_j)| \cdot |x_j^k - x_j^{k^*}|.
\end{aligned}
$$

Define $F^k = (F_{ij}^k)_{(n-1)\times n}$ and $F_{ij}^k = \frac{\partial f_i}{\partial x_j^k}(\xi_j)$. Then we get

$$
\begin{aligned}
\|f(t, x^k) - f(t, x^{k^*})\|_\infty = \|F^k(\xi)(x^k - x^{k^*})\|_\infty &\le \|F^k\|_\infty \cdot \|x^k - x^{k^*}\|_\infty \\
&\le L\|x^k - x^{k^*}\|_\infty
\end{aligned}
$$

where

$$
f(t, x^k) = (f_1(t, x^k), f_2(t, x^k), ..., f_{n-1}(t, x^k))^T
$$

and

$$
L = \max_\xi \|F^k(\xi)\|_\infty = \max_\xi\{\max_i \sum_{l=1}^{n} |\frac{\partial f_i}{\partial x_l^k}(\xi_l)|\} = 4(n-1)\kappa \max_\xi \omega_k(\bar{\xi}) = 4(n-1)\kappa d \bar{\rho}.
$$

Then the local truncation error of x_i^k is expressed as

$$
d_z^i(T_s) = \frac{1}{2}T_s^2 \frac{df_i(t, x^k(t))}{dt} + o(T_s^3) = \frac{1}{2}T_s^2 \frac{df_i(t, x^k(t))}{dt}|_{t=t^*}
$$

where

$$
\begin{aligned}
\frac{df_i(t, x^k(t))}{dt} &= \kappa \sum_{j\in N_i} (\dot{m}_j^k - \dot{m}_i^k) \\
&= \kappa \sum_{j\in N_i} (\omega_k(x_j^k)\dot{x}_j^k - \omega_k(x_{j-1}^k)\dot{x}_{j-1}^k - \omega_k(x_i^k)\dot{x}_i^k + \omega_k(x_{i-1}^k)\dot{x}_{i-1}^k) \\
&\le (n-1)(n+2)\kappa^2\bar{\rho}^2 d^2 l_a.
\end{aligned}
$$

Hence,

$$
\|d_z(T_s)\|_\infty \le \frac{(n-1)(n+2)\kappa^2\bar{\rho}^2 d^2 l_a}{2} T_s^2
$$

where

$$
d_z(T_s) = (d_z^1(T_s), d_z^2(T_s), ..., d_z^{n-1}(T_s))^T.
$$

Let

$$
D = \frac{(n-1)(n+2)\kappa^2\bar{\rho}^2 d^2 l_a}{2},
$$

from Theorem 4.4 in [14], we can conclude that the global truncation error $e_z^k = x^k(t_z) - x^k(z)$ on stripe k satisfies

$$\|e_z^k\| \leq \frac{(n+2)\kappa\bar{\rho}dl_a}{8}(\gamma - 1)T_{s'} + \gamma\|e_0^k\|$$

where $\gamma = e^{\frac{4(n-1)\kappa d^2\bar{\rho}^2 l_a}{\nu}}$. □

Remark 6.2 In the above lemma, we consider general network topologies, and the bound holds for any communication interconnection between agents. For the certain interconnection network, we can get the tighter bound.

Then, we give the key theoretical result in this paper.

Theorem 6.1 *Suppose the initial stripe is partitioned with equal workload. With the discrete time sweep coverage algorithm, the extra time to sweep D spent more than the optimal coverage time is bounded by*

$$\Delta T \leq \frac{dl_a}{\nu}\sqrt{\frac{\beta}{n}}\sum_{j=1}^{q-1} j\alpha^{q-1-j}e^{-\lambda_2\kappa(q-j)\underline{t}} + \frac{(n+2)\kappa d^2\bar{\rho}^2 l_a}{4\nu}\left(\frac{1-\gamma^q}{1-\gamma} - q\right)T_s \tag{6.10}$$

with $\underline{t} = \frac{d\rho l_a}{n\nu}$.

Proof Since agents sweep their respective sub-stripes simultaneously, the time of completing sweeping the stripe depends on the sub-stripe with the most workload. Obviously, optimal partition of the stripe leads to sub-stripes with equal workload. We first estimate the difference between the most workload and average workload on the subregion of each stripe. Then the extra time to sweep each stripe compared to the optimal partition is calculated. Moreover, the error caused by the discretization is also considered. Finally, we obtain the upper bound of the extra time of sweeping D with the proposed discrete time sweep coverage algorithm. From (6.8) in Lemma 6.2, we have

$$H_k(t) \leq H_k^a e^{-2\lambda_2\kappa t}$$

The time used to partition the stripe A_{k+1} is exactly the time to sweep the stripe A_k, which has the lower bound $\underline{t} = \frac{d\rho l_a}{n\nu}$. Therefore, we have

$$H_{k+1}^b \leq H_{k+1}(\underline{t}) \leq H_{k+1}^a e^{-2\lambda_2\kappa\underline{t}}$$

Substituting (6.9) in Lemma 6.3 into the above inequality, we get

$$H_{k+1}^b \leq \left(\alpha^2 H_k^b + \beta\frac{d^2 l_a^2}{n}\right)e^{-2\lambda_2\kappa\underline{t}} \tag{6.11}$$

Thus,

$$\sqrt{H_{k+1}^b} \leq \left(\alpha\sqrt{H_k^b} + dl_a\sqrt{\frac{\beta}{n}}\right) e^{-\lambda_2 \kappa \underline{t}}$$

In addition, since there is no partition operation for the first stripe, $\|e_z^1\| = 0$. Moreover, we have $e_0^{k+1} = e_z^k$ from $x_{k+1}(t_0) = x_k(t_z)$ and $x_{k+1}(0) = x_k(z)$. Therefore, by Lemma 6.4 we get

$$\|e_z^2\| \leq \frac{(n+2)\kappa\bar{\rho}dl_a}{8}(\gamma - 1)T_s,$$

$$\|e_z^3\| \leq \frac{(n+2)\kappa\bar{\rho}dl_a}{8}(\gamma - 1)T_s + \gamma\|e_0^3\| \leq \frac{(n+2)\kappa\bar{\rho}dl_a}{8}(\gamma^2 - 1)T_s$$

Similarly, we obtain the partition error caused by discretization on strip k as follows

$$\|e_z^k\| \leq \frac{(n+2)\kappa\bar{\rho}dl_a}{8}(\gamma^{k-1} - 1)T_s, \quad k = 1, 2, ..., q$$

Since the first stripe is partitioned with equal workload, $H_1^b = 0$ and the extra time to sweep A_1 is 0. Hence, the extra time of sweeping the whole region D is bounded by

$$\begin{aligned}
\Delta T &\leq \frac{1}{\upsilon}\sum_{k=1}^{q}\left(\sqrt{H_k^b} + 2\|e_z^i\|d\bar{\rho}\right) \\
&\leq \frac{dl_a}{\upsilon}\sqrt{\frac{\beta}{n}}\sum_{j=1}^{q-1} j\alpha^{q-1-j}e^{-\lambda_2\kappa(q-j)\underline{t}} + \frac{(n+2)\kappa d^2\bar{\rho}^2 l_a}{4\upsilon}\left(\frac{1-\gamma^q}{1-\gamma} - q\right)T_s.
\end{aligned}$$

which completes the proof of the theorem. □

Intuitively, more communications among agents yield faster convergent rate. However, the statement is not true. Here is a counter-example. Consider two types of interconnection graph with four agents: the directed chain $\mathcal{G}_m$ and the undirected chain $\mathcal{G}_u$. Obviously, communication links in $\mathcal{G}_u$ are more than that in $\mathcal{G}_m$. For simplicity, we set $\rho(x, y) = 1$ and $d = 1$, and so $\sigma_i = \omega_k(x_i^k) = 1, i \in E_n$. Then we have

$$P_u(\mathcal{G}_m) = \begin{pmatrix} 1 & -1 & 0 & 0 \\ -1 & 2 & -1 & 0 \\ 0 & -1 & 2 & -1 \\ 0 & 0 & -1 & 1 \end{pmatrix}$$

and

$$P_u(\mathcal{G}_u) = \begin{pmatrix} 1 & -1.5 & 0.5 & 0 \\ -1.5 & 3 & -2 & 0.5 \\ 0.5 & -2 & 3 & -1.5 \\ 0 & 0.5 & -1.5 & 1 \end{pmatrix}$$

However, $\lambda_2(P_u(\mathcal{G}_m)) = 0.59$ is larger than $\lambda_2(P_u(\mathcal{G}_u)) = 0.17$, which implies that the convergent rate of $\mathcal{G}_m$ outperforms that of $\mathcal{G}_u$. The unusual phenomenon may result from the disparity between consensus object (workload on sub-stripes) and control variable (positions of partition marks), and it is worthy of further investigation.

Remark 6.3 For the directed chain $\mathcal{G}_m$, the tuning parameter κ should be chosen with the constraint

$$\kappa < \frac{1}{2d\bar{\rho}T_s}$$

to avoid collision between partition marks.

6.4 Simulations

In this section, a numerical example has been calculated to verify the above algorithm using Matlab. For simplicity, we consider 4 agents with the interconnection graph $\mathcal{G}_m$ in the rectangular region D. Parameters of D are selected as follows: $l_a = 4$, $l_b = 6$, $d = 1$ (that is $q = 6$), $\kappa = 5$, $\underline{\rho} = 1$, $\bar{\rho} = 2$ and

$$\rho(x, y) = 0.5\sin(x + y) + 1.5$$

In addition, 4 agents with $v = 60$ are used to sweep the region, and sampling period T_s is sufficiently small. Initially, the first stripe has been divided into 4 sub-stripes with equal workload. Then sweeping and partition operations are carried out simultaneously. Colored parts in Fig. 6.3 denote the region that has been swept, and sub-stripes of the same color are covered by the same agent. Finally, the complete coverage of D is finished with the time 0.15, and the optimal coverage time is 0.14. Hence, the extra time to sweep D by multiple agents with interconnection graph $\mathcal{G}_m$ is 0.01, which is less than the upper bound 0.09, given in the theoretical results (6.10).

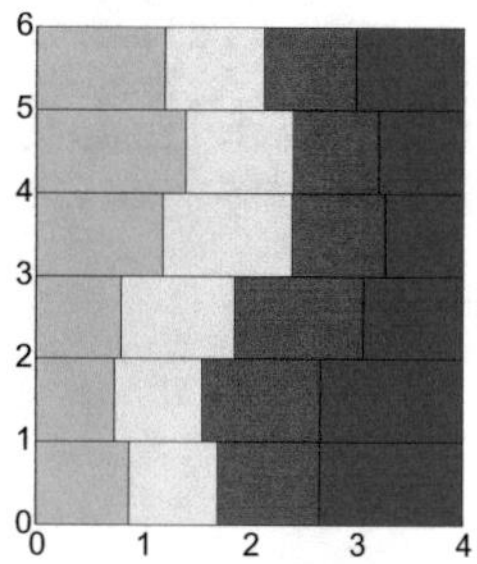

Fig. 6.3 Sweep process of 4 agents using the discrete time sweep coverage algorithm in the region D

6.5 Conclusions

In this chapter, a discrete time formulation was proposed to deal with the sweep coverage problem of multiple agents with general communication topologies in an uncertain region. The theoretical analysis was provided to estimate the upper bound of the coverage time spent more than the optimal time. Moreover, numerical results demonstrate the effectiveness of the discrete time sweep coverage algorithm. However, many problems remain to be solved, including the extension of the sweep coverage algorithm to robotic networks and tighter estimation of coverage time.

References

1. Ren, W., Beard, R.: Distributed Consensus in Multi-vehicle Cooperative Control. Springer, London (2008)
2. Zhang, H., Zhai, C., Chen, Z.: A general alignment repulsion algorithm for flocking of multi-agent systems. IEEE Trans. Autom. Contr. **56**(2), 430–435 (2011)
3. Hong, Y., Hu, J., Gao, L.: Tracking control for multi-agent consensus with an active leader and variable topology. Automatica **42**(7), 1177–1182 (2006)
4. Hu, J., Feng, G.: Distributed tracking control of leader-follower multi-agent systems under noisy measurement. Automatica **46**(8), 1382–1387 (2010)
5. Howard, A., Parker, L.E., Sukhatme, G.: Experiments with a large heterogeneous mobile robot team: Exploration, mapping, deployment and detection. Int. J. Robot. Res. **25**(5–6), 431–447 (2006)
6. Renzaglia, A., Doitsidis, L., Martinelli, A., Kosmatopoulos, E.: Adaptive-based distributed cooperative multi-robot coverage. In: Proceedings of American Control Conference, pp. 486–473. San Francisco (2011)
7. Gage, D.W.: Command control for many-robot systems. In: Proceedings of the 19th Annual AUVS Technical Symposium, pp. 22–24. Huntsville, Alabama (1992)
8. Shi, G., Hong, Y.: Global target aggregation and state agreement of nonlinear multi-agent systems with switching topologies. Automatica **45**(5), 1165–1175 (2009)
9. Cortés, J., Martínez, S., Karatas, T., Bullo, F.: Coverage control for mobile sensing network. IEEE Trans. Robot. Autom. **20**(2), 243–255 (2004)
10. Du, Q., Faber, V., Gunzburger, M.: Centroidal Voronoi tesseuations: applications and algorithms. SIAM Rev. **41**(4), 637–676 (1999)
11. Cheng, T.M., Savkin, A.V.: Decentralized coordinated control of a vehicle network for deployment in sweep coverage. In: Proceedings of the IEEE International Conference on Control and Automation, pp. 275–279. Christchurch, New Zealand (2009)
12. Zhai, C., Hong, Y.: Decentralized sweep coverage algorithm for uncertain region of multi-agent systems. In: Proceedings of American Control Conference. Montréal, Canada (2012)
13. Horn, R.A., Johnson, C.R.: Matrix Analysis. Cambridge University Press (1987)
14. Gear, C.W.: Numerical Initial Value Problems for Ordinary Differential Equations. Prentice-Hall, Englewood Cliffs (1971)

Chapter 7
Coverage-Based Cooperative Interception Against Supersonic Flight Vehicles

7.1 Introduction

The development of modern military technologies has resulted in more complicated battlefields in the past decades. In particular, the manipulability and intelligence of flight vehicles have been improved tremendously. For instance, the flight vehicles may opt to release some decoys in order to increase the possibility of penetration, and one single interceptor fails to defend such flight vehicles (termed as targets) effectively. For this reason, the cooperative interception problem of multiple interceptors has become an important research topic in recent years [1–3].

Due to the measurement noise and decoys, the accurate state of the real target is not available in the real scenario. Nevertheless, the location of the target can be described by a probability density function. There are mainly three control approaches to the guidance of interceptors based on the probability density function of the target: the minimum mean square error criterion (MMSE), the maximum a-posterior probability criterion (MAP) and the highest probability interval criterion (HPI) [4]. The first criterion, MMSE, is to construct a virtual target and then guide the flight vehicles based on the measurement information on the target. The MAP is to choose the point that corresponds to the maximum value of the posterior probability density function as the position of the target. In the HPI, the interceptor is steered towards the predicted position of the target to maximize the probability of the target remaining reachable. A novel predictive guidance law was developed in [5], which extended the work of [4] by introducing the piecewise constant control command and involving multiple decoys. The engagement of multi-interceptor to one highly maneuverable target was studied in [3], and a cooperative guidance method based on the probability information of the target was developed by using the receding horizon optimization. Essentially, these three criteria are the same if the probability density function is strictly concave and symmetric around its only mode [4].

On the other hand, cooperative control has received significant attentions in the aeronautics and astronautics field [6]. A comprehensive survey is presented in [7]

C. Zhai et al., *Cooperative Coverage Control of Multi-Agent Systems and its Applications*, Studies in Systems, Decision and Control 408,
https://doi.org/10.1007/978-981-16-7625-3_7

to address basic problems of multi-agent cooperation and coordination. As a branch of cooperative control, cooperative coverage has been extensively investigated [8–11]. Cooperative coverage of multiple agents derives from many practical problems such as search and rescue, exploration, surveillance, surface-to-air-missile interception and environmental monitor [12–14]. Different optimization formulations have been proposed for coverage problem [7, 15]. The cooperative surveillance problem using a set of heterogeneous unmanned vehicles was considered in [9], and the algorithm for the computation of trajectories was proposed to maximize the spatio-temporal coverage. Reference [10] considered the sweep coverage in the uncertain environment and provided a decentralized coverage algorithm, while [16] designed an integral-based control law for effective coverage problems. This chapter considers how multiple interceptors cooperatively intercept a high maneuvering target with decoys. The contributions of this study are two-fold:

1. A coverage-based approach is proposed to deal with the cooperative interception problem. The interception algorithm consists of cooperative guidance algorithm, which generates desired position of each interceptor and optimal control law, which drives the interceptor to the desired position with minimum energy.
2. The joint probability of interception is estimated under appropriate assumptions. Theoretical analysis demonstrates that the lower bound of the joint interception probability goes up with the increase of the number of interceptors and its lethal radius. This implies that the number of interceptors can compensate for its poor maneuverability in order to guarantee the desirable probability of interception.

This chapter will proceed as follows: the coverage-based interception problem is formulated in Sect. 7.2. Then the interception algorithm and relevant analysis are presented in Sect. 7.3. Simulation results are provided to validate the proposed interception algorithm in Sect. 7.4. Finally, the conclusions and future directions are given in Sect. 7.5.

7.2 Problem Formulation

Consider the planar interception problem of multiple interceptors against the target vehicle with decoys (see Fig. 7.1). Notice that the horizontal direction corresponds to x-axis, and the vertical direction corresponds to y-axis. The real target to be intercepted is the high supersonic flight vehicle, which is able to release decoys in order to confuse the interceptors during flight. The interceptors are equipped with discriminators to identify the real target. Due to the interference of decoys, the high maneuvering target appears as more than one apparent target to the interceptors during the engagement, and the decoys are also thought of as targets before they are discriminated. In addition, each interceptor has the finite lethal radius, and the interception is achieved successfully if the miss distance between the interceptor and the target is within the lethal radius of the interceptor. The cooperation among multiple interceptors contributes to sharing more information on the target and increasing the

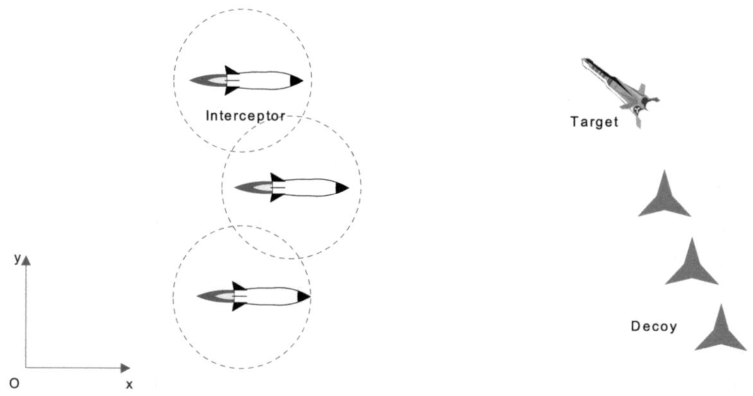

Fig. 7.1 Planar engagement geometry. The circle around each interceptor indicates the lethal radius of the interceptor

interception probability. The aim of this study is to develop the cooperative interception algorithm to maximize the interception probability of the real target.

Let $x_I^i \in R^2$ and $x_T^j \in R^2$ denote the position of the ith interceptor and that of the jth target, respectively. Then the kinematics of each interceptor is given by

$$\dot{x}_I^i = v_I^i, \quad \dot{v}_I^i = u_i, \quad i \in E_n = \{1, 2, ..., n\}$$

where u_i denotes the bounded control signal, i.e., $\|u_i(t)\| \le u_{max}$. Let $X_I^i = [x_I^{i\,T} \; v_I^{i\,T}]^T$, and then the above equation can be expressed as a canonical linear time-invariant system

$$\dot{X}_I^i = A_I X_I^i + B_I u_i, \tag{7.1}$$

where

$$A_I = \begin{bmatrix} 0 & 0 & 1 & 0 \\ 0 & 0 & 0 & 1 \\ 0 & 0 & 0 & 0 \\ 0 & 0 & 0 & 0 \end{bmatrix}, \quad B_I = \begin{bmatrix} 0 & 0 \\ 0 & 0 \\ 1 & 0 \\ 0 & 1 \end{bmatrix}.$$

Similarly, the motion equation of the jth target is expressed as

$$\dot{x}_T^j = v_T^j, \quad \dot{v}_T^j = a_j, \quad j \in E_m = \{1, 2, ..., m\}$$

where the acceleration a_j is bounded, i.e., $\|a_j\| \le a_{\max}$, and it is not available to the interceptors. Assume the acceleration of the target a_j is a Wiener process, or more precisely, the acceleration is a process with independent increments, which is not necessarily a Wiener process [17] (i.e., $\dot{a}_j(t) = w(t)$ and $w(t)$ is a white noise process). The kinematic model of the target can be augmented if other maneuver forms are adopted, for instance, the constant velocity maneuver. Moreover, each interceptor equipped with sensors can get the discrete measurement on the state of targets as follows:

$$\psi(t_k) = h(t_k, \xi), \quad t_k = k\Delta t, \quad k = 1, 2, ..., N$$

where Δt represents the sampling time interval, and the noise ξ is assumed to be normally distributed. Suppose that state equations and output equations of the system satisfy the observability condition such that the state of targets can be reconstructed from the measurements.

Since the interceptors and targets are approaching along the horizontal direction, it is crucial to determine the positions of targets along the vertical direction in order to intercept the targets. Considering that accurate positions of targets along the vertical direction are not available, a density function $f_j(\tau)$ is constructed to describe the probability of the jth target occurring at τ. Here, $f_j(\tau)$ satisfies $f_j(\tau) \geq 0, \forall \tau \in R$ and

$$\int_{-\infty}^{+\infty} f_j(\tau)d\tau = 1.$$

In particular, the probability density function can be computed by employing the Kalman filter, which allows to update the mean value and the covariance of the lateral positions of targets with the measurement $\psi(t_k)$. Thus, the probability density function of the jth target at time t_k is given by,

$$f_j(\tau) = \frac{1}{\sqrt{2\pi}\sigma_j} e^{-\frac{(\tau-\mu_j)^2}{2\sigma_j^2}}.$$

where μ_j and σ_j denote the expectation and standard error of the distribution, respectively. In addition, $\sigma_j \leq \sigma_{max}, \forall j \in E_m$ and σ_{max} is a known parameter, which characterizes the measurement precision of sensors. Moreover, the probability density function of each target is available to all the interceptors. Each target is also associated with a probability at time t_k, denoted by $\eta_j(t_k)$, $j \in E_m$ with constraints $0 \leq \eta_j(t_k) \leq 1$ and $\sum_{j=1}^{m} \eta_j(t_k) = 1$, which is the probability that the target is the real one. The probabilities are provided by the discriminator equipped on interceptors. Then the probability density function of the real target along the vertical direction at time t_k is presented as follows

$$F_k(\tau) = \sum_{j=1}^{m} \eta_j(t_k) f_j(\tau) \tag{7.2}$$

Let t_f represent the total flight time of the interceptors during the engagement, and it depends on the approaching speed and the initial relative range between the interceptors and the targets. Because the interceptors are launched simultaneously, suppose the total flight time of interceptors are the same and known to each interceptor.

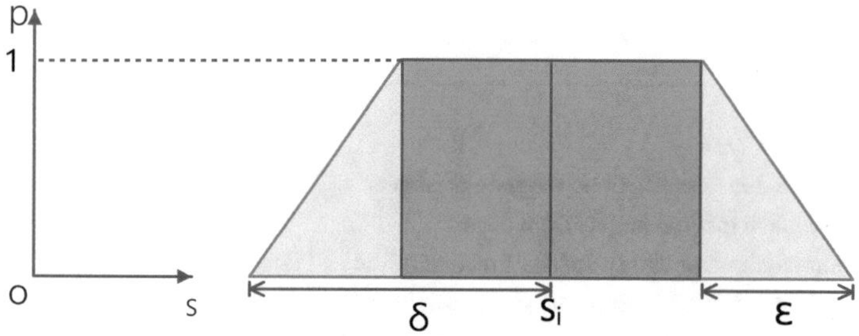

Fig. 7.2 Interception probability of the ith interceptor along the vertical direction

Inspired by the work in [11], the joint interception probability of n cooperative interceptors at time t_k is defined as

$$P_k(s) = \int_{-\infty}^{+\infty} \left[1 - \prod_{i=1}^{n} (1 - p(\tau, s_i)) \right] F_k(\tau) d\tau \tag{7.3}$$

where $s = (s_1, s_2, ..., s_n) \in R^n$, $s_i \triangleq D_I x_I^i$, $D_I = [0 \;\; 1]$ and

$$p(\tau, s_i) = \begin{cases} 1, & |\tau - s_i| \leq \delta - \epsilon; \\ 0, & |\tau - s_i| \geq \delta; \\ -\frac{1}{\epsilon}(\tau - s_i) + \frac{\delta}{\epsilon}, & \delta - \epsilon < \tau - s_i < \delta; \\ \frac{1}{\epsilon}(\tau - s_i) + \frac{\delta}{\epsilon}, & -\delta < \tau - s_i < \epsilon - \delta. \end{cases}$$

describes the relationship between the interception probability of the ith interceptor and the distance between this interceptor and the target (see Fig. 7.2). Here δ denotes the lethal radius of interceptors. The target will be destroyed if the distance between the target and one of interceptors is less than δ. In particular, ϵ is used to avoid the discontinuity of $p(\tau, s_i)$, and it can be tuned sufficiently small. Essentially, $P_k(s)$ represents the probability that the real target can be intercepted by at least one interceptor. Let $R(t_{k+1}, x_I^i(t_k))$ denote the reachable set of the ith interceptor at time t_{k+1} with the condition that its position at time t_k is $x_I^i(t_k)$. The range of $R(t_{k+1}, x_I^i(t_k))$ along the vertical direction can be computed according to the formula in [5]. The objective of this study is to develop the cooperative interception algorithm in order to solve the following optimization problem

$$\max_{s \in R^n} P_k(s)$$

at the final time t_f.

The interception procedure is shown in Table 7.1. When new measurements on the state of targets are available at time t_k, the interceptors distinguish the real target from decoys with the discriminator and compute the probability density function of the real target by a Kalman filter. Then each interceptor implements the cooperative

Table 7.1 Coverage based interception algorithm

```
set k = 1
while (time < t_f)
    Obtain the measurement on the state of targets
    Distinguish the real targets from decoys
    Compute the probability density function of the real target (7.2)
    Obtain the desired position at t_{k+1} by implementing (7.5)
    Move towards the desired position under optimal control (7.6)
    k = k + 1
end while
```

guidance algorithm and gets its desired position at time t_{k+1}. If the desired position is not within the reachable set, choose the nearest point in the reachable set as the desired position. Next, each interceptor moves towards its desired position under the optimal control command. Finally, the above processes are repeated until the total flight time is achieved.

The following assumptions are presented to facilitate the analysis.

- Each interceptor is connected to its nearest neighbors along the vertical direction.
- The real target can be distinguished by each interceptor before the engagement.

Considering that both the acceleration of targets and the sampling period of the sensor equipped on the interceptor are finite, there exists the upper bound on the difference between the actual peak and the estimated peak of the probability density function for each target. Each interceptor can compute the velocity of the target according to the measurements, which include the rate of the line of sight angle per time unit and the relative distance. Without allowing for the uncertainty of the current state estimate, the upper bound γ can be calculated as follows

$$\gamma = \frac{a_{max}}{2}\Delta t^2$$

7.3 Main Results

In this section, theoretical results on the cooperative interception algorithm are presented to estimate the joint interception probability of multiple interceptors. Construct the following objective function

$$H_k(s) = \theta \underbrace{(1 - P_k(s))}_{\text{Coverage}} + \underbrace{(1-\theta)\sum_{i=1}^{n-1}(s_{i+1} - s_i - 2(\delta - \lambda))^2}_{\text{Formation}} \tag{7.4}$$

where

$$\lambda = \epsilon + \frac{\theta(n-2)}{4\epsilon(1-\theta)}$$

and it satisfies $\lambda < \delta$. In addition, $\theta \in (0, 1)$ is a tunable parameter, and it is close to 0 so that λ is approximate to ϵ. $H_k(s)$ is composed of the "coverage" term and the "formation" term. The aim is to maximize $P_k(s)$ and meanwhile link together the lethal radius of interceptors to form an "interception chain". Thus, the desired configuration of interceptors is achieved if both terms converge to 0. To obtain the desired position, each interceptor carries out the following coverage-based guidance algorithm

$$\dot{s}_i = -\kappa \frac{\partial H_k}{\partial s_i}, \quad i \in E_n \tag{7.5}$$

where κ is a positive constant.

Theorem 7.1 *With the cooperative guidance algorithm (7.5), collisions between interceptors can be avoided.*

Proof For two interceptors i and $i+1$ with lateral positions s_i and s_{i+1} ($s_{i+1} > s_i$), the time derivatives of s_i and s_{i+1} are given by

$$\begin{aligned}
\dot{s}_i &= -\kappa \frac{\partial H_k}{\partial s_i} \\
&= \kappa\theta \frac{\partial P_k}{\partial s_i} - \kappa(1-\theta)\frac{\partial}{\partial s_i}\sum_{j=1}^{n-1}(s_{j+1} - s_j - 2(\delta - \lambda))^2 \\
&= \kappa\theta \int_{-\infty}^{+\infty} \frac{\partial(1 - \prod_{i=1}^{n}(1 - p(\tau, s_i)))}{\partial s_i} F_k(\tau)d\tau \\
&\quad - \kappa(1-\theta)\frac{\partial}{\partial s_i}\sum_{j=1}^{n-1}(s_{j+1} - s_j - 2(\delta - \lambda))^2 \\
&= \kappa\theta \int_{s_i-\delta}^{s_i+\delta} \frac{\partial p(\tau, s_i)}{\partial s_i} \prod_{j=1, j\neq i}^{n} (1 - p(\tau, s_j))F_k(\tau)d\tau \\
&\quad - \kappa(1-\theta)\frac{\partial}{\partial s_i}\sum_{j=1}^{n-1}(s_{j+1} - s_j - 2(\delta - \lambda))^2 \\
&= -\frac{\kappa\theta}{\epsilon} \int_{s_i-\delta}^{s_i-\delta+\epsilon} \prod_{j=1, j\neq i}^{n} (1 - p(\tau, s_j))F_k(\tau)d\tau \\
&\quad + \frac{\kappa\theta}{\epsilon} \int_{s_i+\delta-\epsilon}^{s_i+\delta} \prod_{j=1, j\neq i}^{n} (1 - p(\tau, s_j))F_k(\tau)d\tau \\
&\quad - \kappa(1-\theta)\frac{\partial}{\partial s_i}\sum_{j=1}^{n-1}(s_{j+1} - s_j - 2(\delta - \lambda))^2
\end{aligned}$$

and

$$
\begin{aligned}
\dot{s}_{i+1} &= -\kappa \frac{\partial H_k}{\partial s_{i+1}} \\
&= -\frac{\kappa\theta}{\epsilon} \int_{s_{i+1}-\delta}^{s_{i+1}-\delta+\epsilon} \prod_{j=1, j\neq i+1}^{n} (1-p(\tau, s_j)) F_k(\tau) d\tau \\
&\quad + \frac{\kappa\theta}{\epsilon} \int_{s_{i+1}+\delta-\epsilon}^{s_{i+1}+\delta} \prod_{j=1, j\neq i+1}^{n} (1-p(\tau, s_j)) F_k(\tau) d\tau \\
&\quad - \kappa(1-\theta) \frac{\partial}{\partial s_{i+1}} \sum_{j=1}^{n-1} (s_{j+1} - s_j - 2(\delta - \lambda))^2
\end{aligned}
$$

When $0 \leq s_{i+1} - s_i \leq 2(\delta - \epsilon)$, it follows from the definition of $p(\tau, s_i)$ that

$$
p(\tau, s_i) = 1, \quad s_{i+1} - \delta \leq \tau \leq s_{i+1} - \delta + \epsilon
$$

and

$$
p(\tau, s_{i+1}) = 1, \quad s_i + \delta - \epsilon \leq \tau \leq s_i + \delta
$$

The time derivatives of s_i and s_{i+1} can be simplified as follows

$$
\begin{aligned}
\dot{s}_i &= -\frac{\kappa\theta}{\epsilon} \int_{s_i-\delta}^{s_i-\delta+\epsilon} \prod_{j=1, j\neq i}^{n} (1-p(\tau, s_j)) F_k(\tau) d\tau \\
&\quad - \kappa(1-\theta) \frac{\partial}{\partial s_i} \sum_{j=1}^{n-1} (s_{j+1} - s_j - 2(\delta - \lambda))^2
\end{aligned}
$$

and

$$
\begin{aligned}
\dot{s}_{i+1} &= \frac{\kappa\theta}{\epsilon} \int_{s_{i+1}+\delta-\epsilon}^{s_{i+1}+\delta} \prod_{j=1, j\neq i+1}^{n} (1-p(\tau, s_j)) F_k(\tau) d\tau \\
&\quad - \kappa(1-\theta) \frac{\partial}{\partial s_{i+1}} \sum_{j=1}^{n-1} (s_{j+1} - s_j - 2(\delta - \lambda))^2
\end{aligned}
$$

Let $d_i = s_{i+1} - s_i$, $1 \leq i \leq (n-1)$, and then

$$
\begin{aligned}
\dot{d}_1 &= \frac{\kappa\theta}{\epsilon} \int_{s_2+\delta-\epsilon}^{s_2+\delta} \prod_{j=1, j\neq 2}^{n} (1-p(\tau, s_j)) F_k(\tau) d\tau \\
&\quad + \frac{\kappa\theta}{\epsilon} \int_{s_1-\delta}^{s_1-\delta+\epsilon} F_k(\tau) d\tau \\
&\quad + 2\kappa(1-\theta)(d_2 + 2(\delta - \lambda) - 2d_1)
\end{aligned}
$$

$$\dot{d}_i = \frac{\kappa\theta}{\epsilon}\int_{s_{i+1}+\delta-\epsilon}^{s_{i+1}+\delta} \prod_{j=1, j\neq i+1}^{n} (1 - p(\tau, s_j))F_k(\tau)d\tau$$
$$+ \frac{\kappa\theta}{\epsilon}\int_{s_i-\delta}^{s_i-\delta+\epsilon} \prod_{j=1, j\neq i}^{n} (1 - p(\tau, s_j))F_k(\tau)d\tau$$
$$+ 2\kappa(1-\theta)(d_{i+1} + d_{i-1} - 2d_i),\ 2 \leq i \leq (n-2)$$

$$\dot{d}_{n-1} = \frac{\kappa\theta}{\epsilon}\int_{s_{n-1}-\delta}^{s_{n-1}-\delta+\epsilon} \prod_{j=1, j\neq n-1}^{n} (1 - p(\tau, s_j))F_k(\tau)d\tau$$
$$+ \frac{\kappa\theta}{\epsilon}\int_{s_n+\delta-\epsilon}^{s_n+\delta} F_k(\tau)d\tau$$
$$+ 2\kappa(1-\theta)(d_{n-2} + 2(\delta - \lambda) - 2d_{n-1})$$

$d_1 \leq (\delta - \lambda)$ results in $\dot{d}_1 > 0$, which implies that interceptor 1 and interceptor 2 will not get close to each other any more when the distance between them is less than $\delta - \lambda$. Similarly, when $d_i \leq (\delta - \lambda)/2^{i-1}$, $2 \leq i \leq (n-2)$ and $d_{n-1} \leq (\delta - \lambda)$, it follows that $\dot{d}_i > 0$, $2 \leq i \leq (n-1)$. Hence, collisions can be avoided with the cooperative guidance algorithm. □

Remark 7.1 This proposition implies that the order of interceptors along the vertical direction remains unchanged when the cooperative guidance algorithm is implemented. Thus, under the control command, the interceptors will move toward their respective desired positions with collision avoidance.

In what follows, consider the design of the optimal control command of each interceptor (see Fig. 7.3). Since the desired position of each interceptor at t_{k+1} is available after implementing the cooperative guidance algorithm, the control design for the system (7.1) can be converted into linear quadratic tracking problem. Construct the performance index in the time interval $[t_k, t_{k+1}]$

$$J_i(t_{k+1}) = \frac{1}{2}e_I^i(t_{k+1})^T Q_0 e_I^i(t_{k+1}) + \int_{t_k}^{t_{k+1}} [e_I^i(t)^T Q_1 e_I^i(t) + u_i(t)^T R u_i(t)]dt,\ i \in E_n$$

where

$$e_I^i(t) = z_i^*(t_{k+1}) - CX_I^i(t)$$

and

$$C = \begin{pmatrix} 1\,0\,0\,0 \\ 0\,1\,0\,0 \end{pmatrix}.$$

Here $z_i^*(t_{k+1}) = [q^*(t_{k+1}), s_i^*(t_{k+1})]^T$ represents the desired position of interceptor i at t_{k+1}. Q_0 and Q_1 are symmetric positive semi-definite, and R is symmetric positive definite. It is easy to obtain the optimal control [18]

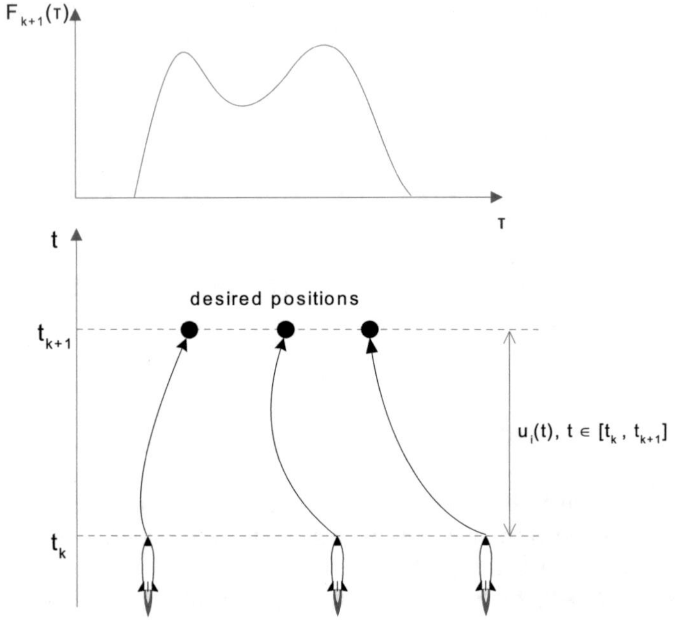

Fig. 7.3 Motion planning and control of interceptors in each sampling period. The upper panel is the joint probability density function of the targets at time t_{k+1}, and the lower panel describes the trajectories of interceptors under optimal control command

$$u_i(t) = -R^{-1}B_I^T K(t)X_I^i + R^{-1}B_I^T g(t) \tag{7.6}$$

where $K(t)$ satisfies the continuous time Riccati equation

$$\dot{K}(t) = -K(t)A_I - A_I^T K(t) + K(t)B_I R^{-1}B_I^T K(t) - C^T Q_1 C \tag{7.7}$$

with the terminal condition

$$K(t_{k+1}) = C^T Q_0 C.$$

Likewise, $g(t)$ satisfies the non-homogeneous vector differential equation

$$\dot{g}(t) = -(A_I - B_I R^{-1} K(t))g(t) - C^T Q_1 z_i^*(t_{k+1}) \tag{7.8}$$

with the final condition

$$g(t_{k+1}) = C^T Q_0 z_i^*(t_{k+1}).$$

Remark 7.2 $s_i^*(t_{k+1})$ can be calculated according to the cooperative guidance algorithm (7.5). In addition, all the interceptors have the same horizontal coordinate of the desired position $q^*(t_{k+1})$, which depends on the horizontal speed of interceptors and is specified in advance. Riccati equation (7.7) and vector differential equation

(7.8) can be solved backwards using the boundary conditions. In practice, matrices Q_0, Q_1 and R should be tuned properly such that $\|e_I^i(t_{k+1})\|$ tends to 0 as much as possible.

Then, the main result on theoretical analysis is presented as follows.

Theorem 7.2 *Suppose that the two assumptions hold and the desired positions of interceptors at the final time are within their respective reachable set. With the cooperative guidance algorithm (7.5) and optimal control law (7.6), the joint interception probability of n interceptors is bounded by*

$$P \geq \frac{1}{\sqrt{2\pi}\sigma_{max}} \int_{-(\delta-\lambda)n-\gamma}^{(\delta-\lambda)n-\gamma} e^{-\frac{\tau^2}{2\sigma_{max}^2}} d\tau$$

where

$$\lambda = \epsilon + \frac{\theta(n-2)}{4\epsilon(1-\theta)}$$

Proof When the real target is distinguished by each interceptor before the engagement, the joint probability density function of the targets becomes $f_j(\tau)$ for some $j \in E_m$. Thus, the desired positions of interceptors at the final time can be calculated by solving the equations

$$\frac{\partial H_k}{\partial s_i} = 0, \quad i \in E_n$$

which is equivalent to the following equations.

$$\begin{aligned} &\theta \frac{\partial P_k}{\partial s_1} + 2(1-\theta)(s_2 - s_1 - 2(\delta - \lambda)) = 0 \\ &\theta \frac{\partial P_k}{\partial s_i} - 2(1-\theta)(2s_i - s_{i-1} - s_{i+1}) = 0, \ 2 \leq i \leq (n-1) \\ &\theta \frac{\partial P_k}{\partial s_n} - 2(1-\theta)(s_n - s_{n-1} - 2(\delta - \lambda)) = 0 \end{aligned} \tag{7.9}$$

The inequality

$$-\frac{1}{\epsilon} \leq \frac{\partial P_k}{\partial s_i} \leq \frac{1}{\epsilon}, \ (1 \leq i \leq n)$$

gives

$$2(\delta - \lambda) - \frac{\theta(i-1)}{2\epsilon(1-\theta)} \leq s_i - s_{i-1} \leq 2(\delta - \lambda) + \frac{\theta(i-1)}{2\epsilon(1-\theta)}$$

for $2 \leq i \leq (n-1)$ and

$$2(\delta - \lambda) - \frac{\theta}{2\epsilon(1-\theta)} \leq s_n - s_{n-1} \leq 2(\delta - \lambda) + \frac{\theta}{2\epsilon(1-\theta)}$$

Take

$$\lambda = \epsilon + \frac{\theta(n-2)}{4\epsilon(1-\theta)},$$

and then it leads to

$$s_i - s_{i-1} \leq 2(\delta - \epsilon), \ \forall\, 2 \leq i \leq n$$

which implies that the lethal radiuses of all interceptors are linked together to form an "interception chain". Accumulating all the equations in (7.9) yields

$$\sum_{i=1}^{n} \frac{\partial P_k}{\partial s_i} = 0$$

which is equivalent to

$$\int_{s_1-\delta}^{s_1-\delta+\epsilon} f_j(\tau)d\tau = \int_{s_n+\delta-\epsilon}^{s_n+\delta} f_j(\tau)d\tau$$

since $F_k(\tau) = f_j(\tau)$ at the last sampling period. It follows from the symmetry of normal distribution that $\mu_j - s_1 = s_n - \mu_j$, and this means that the above "interception chain" is symmetric with respect to μ_j.

Considering that the difference between the actual peak and the estimated peak of the probability density functions is bounded by γ, the probability of interception at the final moment satisfies

$$\begin{aligned} P &\geq \int_{\mu_j-(\delta-\lambda)n}^{\mu_j+(\delta-\lambda)n} f_j(\tau-\gamma)d\tau \\ &= \frac{1}{\sqrt{2\pi}\sigma_j} \int_{\mu_j-(\delta-\lambda)n}^{\mu_j+(\delta-\lambda)n} e^{-\frac{(\tau-\mu_j-\gamma)^2}{2\sigma_j^2}} d\tau \\ &\geq \frac{1}{\sqrt{2\pi}\sigma_{max}} \int_{\mu_j-(\delta-\lambda)n}^{\mu_j+(\delta-\lambda)n} e^{-\frac{(\tau-\mu_j-\gamma)^2}{2\sigma_{max}^2}} d\tau \\ &= \frac{1}{\sqrt{2\pi}\sigma_{max}} \int_{-(\delta-\lambda)n-\gamma}^{(\delta-\lambda)n-\gamma} e^{-\frac{\tau^2}{2\sigma_{max}^2}} d\tau \end{aligned}$$

which completes the proof. □

Remark 7.3 According to the theoretical result, the lower bound of the joint probability of interception goes up with the increase of the number of interceptors n and the lethal radius δ.

Remark 7.4 To guarantee the accuracy of the estimation for the probability of interception, θ and ϵ should be tuned properly such that

$$\frac{\theta(n^2 - 5n + 8)}{4\epsilon(1 - \theta)}$$

is sufficiently small.

7.4 Numerical Simulation

In this section, simulation results are provided to verify the cooperative interception algorithm and theoretical results. In order to illustrate the dynamic process of the interception, take 3 interceptors cooperatively intercepting a real target with a decoy, and the decoy is discriminated by each interceptor before the engagement. At the initial time, the velocities of interceptors and targets are assumed to be aligned with the line of sight. The positions of the real target and the decoy are 2 and 7 and the standard deviation of the measurement noise σ_{max} is set to be 1. According to the interception algorithm, the probability of interception mainly depends on the state of targets and interceptors in the final time interval $[t_f - \Delta t, t_f]$ regardless of their trajectories. For simplicity, both the real target and the decoy move straight along the horizontal direction. For high maneuverability targets, the final outcome should remain unchanged if the desired positions of targets are located within the reachable set of the interceptors in the final time interval. The initial positions of 3 interceptors in the vertical direction are 3, 6 and 8, respectively. The lethal radius of the intercepting flight vehicle δ is 1 and ϵ is set to be 0.5. The total flight time t_f is $10s$, and $\eta_1 = \eta_2 = 0.5$. Assume that at $t = 5s$ the real target is discriminated and $\eta_1 = 1$, $\eta_2 = 0$. Figure 7.4 presents the trajectories of 3 interceptors and the targets. 3 interceptors get closer to each other in order to allow for the coverage level as θ goes up to 0.2 and 0.4. In comparison, theses interceptors come together first and then move apart when θ increases to 0.6, 0.8. This is because the condition $\lambda < \delta$ does not hold any more when θ exceeds 0.5.

In addition, extensive simulations have been carried out to verify the theoretical bound of the joint interception probability. The simulation parameters are given as follows: the standard deviation of the measurement noise $\sigma_{max} = 1$, and the measurement frequency of each interceptor is 5 Hz (i.e., $\Delta t = 0.2s$). The total flight time $t_f = 10s$, $\kappa = 2 \times 10^3$, $a_{max} = 5$, $\theta = 0.01$, $\eta_1 = \eta_2 = 0.5$, the lethal radius of the interceptor δ ranges from 0.6 to 1.4 ($\epsilon = \frac{\delta}{2}$). In Fig. 7.5a, 3 interceptors cooperatively intercept the targets (one real target and one decoy). The probability of interception goes up with the increase of the lethal radius δ for both numerical and theoretical results, and the theoretical bound becomes tighter for the larger lethal radius. The similar result appears for the number of interceptors with the constant lethal radius $\delta = 0.2$ in Fig. 7.5b. Notably, the probability of interception gets close to 1 as the number of interceptors increases to 5. Finally, it is worth pointing out that each numerical data point in Fig. 7.5 is obtained by repeating the interception scenario 100 times, and each simulation has the different noise disturbance and initial conditions.

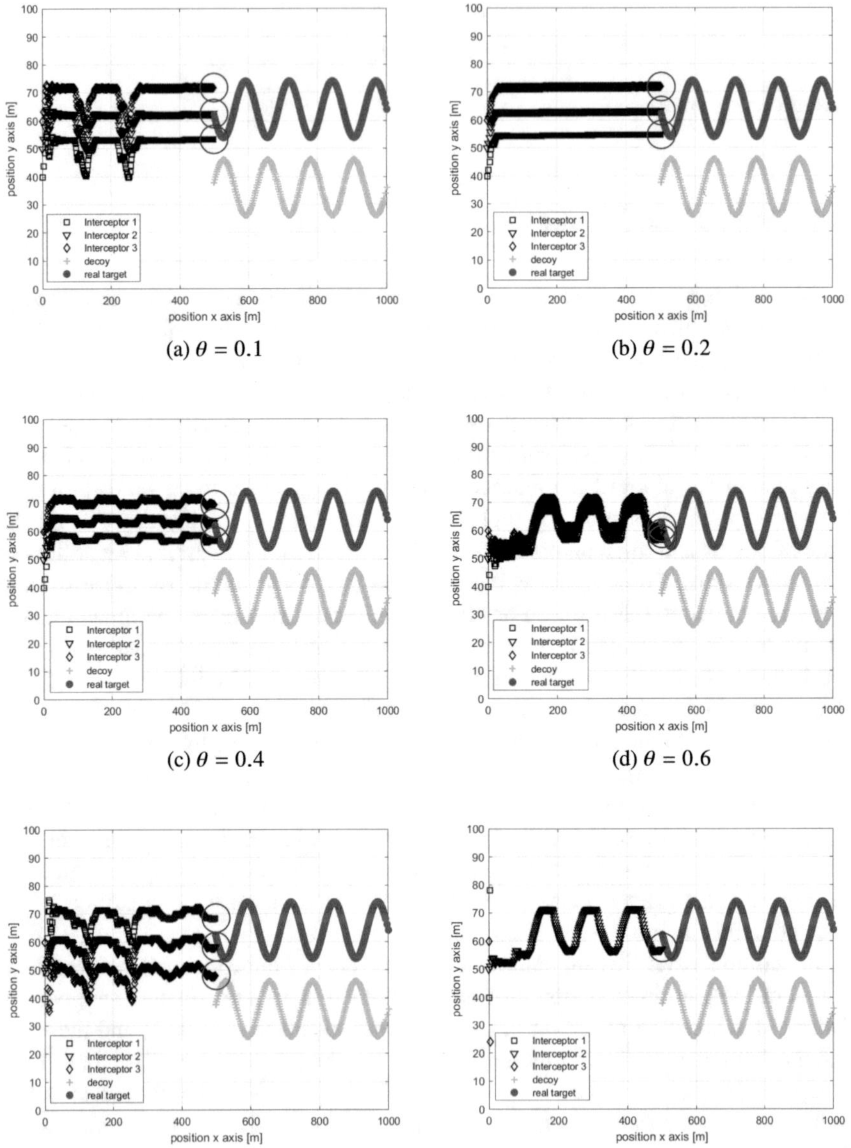

(a) $\theta = 0.1$ (b) $\theta = 0.2$

(c) $\theta = 0.4$ (d) $\theta = 0.6$

(e) $\theta = 0.8$ (f) $\theta = 0.95$

Fig. 7.4 Trajectories of the intercepting flight vehicles and the targets

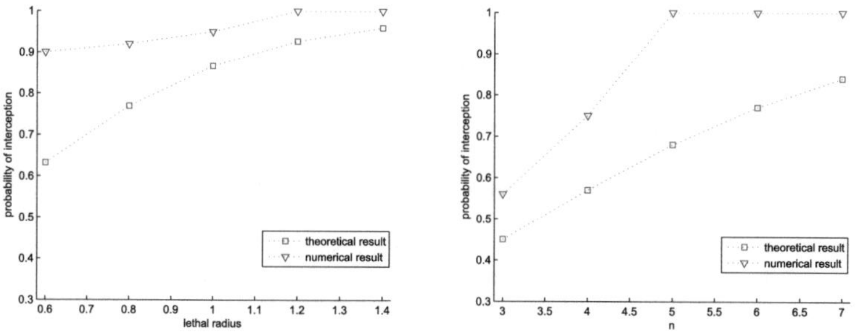

(a) number of interceptors: $n = 3$ (b) lethal radius: $\delta = 0.2$

Fig. 7.5 The joint interception probability with respect to the lethal radius (**a**) and the number of interceptors (**b**) for a mobile target and a decoy

7.5 Conclusions

By integrating the coverage-based guidance algorithm with the optimal control law, the cooperative interception algorithm was proposed for multiple interceptors to maximize the probability of destroying the target vehicle with decoys. Moreover, the lower bound of the interception probability was estimated under appropriate assumptions, and theoretical results were validated by numerical simulations. Future work may include the relaxation of the present assumptions and the consideration of non-holonomic constraints on interceptors.

References

1. Jeon, I., Lee, J., Tahk, M.: Homing guidance law for cooperative attack of multiple missiles. J. Guid. Control. Dyn. **33**(1), 275–280 (2010)
2. Ben-Asher, J.: Minimum-effort interception of multiple targets. J. Guid. Control. Dyn. **16**(3), 600–602 (1993)
3. Yang, B., Liu, H., Yao, Y.: Cooperative interception guidance for multiple vehicles: a receding horizon optimization approach. In: Proceedings of 2014 IEEE Chinese Guidance, Navigation and Control Conference, pp. 827–831. IEEE, Yantai (2014)
4. Dionne, D., Michalska, H., Rabbath, C.: A predictive guidance law with uncertain information about the target state. In: 2006 American Control Conference, pp. 1062–1067. IEEE (2006)
5. Dionne, D., Michalska, H., Rabbath, C.: Predictive guidance for pursuit-evasion engagements involving multiple decoys. J. Guid. Control. Dyn. **30**(5), 1277–1286 (2007)
6. Ben-Asher, Y., et al.: Distributed decision and control for cooperative UAVs using ad hoc communication. IEEE Trans. Control Syst. Technol. **16**(3), 511–516 (2008)
7. Hong, Y., Zhai, C.: Dynamic coordination and distributed control design of multi-agent systems. Control Theory Appl. **28**(10), 1506–1512 (2011)
8. Cortés, J., Martínez, S., Karatas, T., Bullo, F.: Coverage control for mobile sensing network. IEEE Trans. Robot. Autom. **20**(2), 243–255 (2004)

9. Ahmadzadeh, A., Jadbabaie, A., Kumar, V., Pappas, G.: Cooperative coverage using receding horizon control. In: Proceedings of European Control Conference, pp. 2466–2470. Kos, Greece (2007)
10. Zhai, C., Hong, Y.: Decentralized sweep coverage algorithm for multi-agent systems with workload uncertainties. Automatica **49**(7), 2154–2159 (2013)
11. Cassandras, G., Li, W.: Sensor networks and cooperative control. Eur. J. Control. **11**(4), 436–463 (2005)
12. Howard, A., Parker, L., Sukhatme, G.: Experiments with a large heterogeneous mobile robot team: exploration, mapping, deployment and detection. Int. J. Robot. Res. **25**(5–6), 431–447 (2006)
13. Zhai, C., Hong, Y.: Decentralized algorithm for online workload partition of multi-agent systems. In: Proceedings of Chinese Control Conference, pp. 4920–4925. IEEE, Yantai (2011)
14. Hu, J., Xie, L., Lum, K., Xu, J.: Multiagent information fusion and cooperative control in target search. IEEE Trans. Control Syst. Technol. **21**(4), 1223–1235 (2012)
15. Hong, Y., Zhang, Y.: Distributed optimization: algorithm design and convergence analysis. Control Theory Appl. **31**(7), 850–857 (2014)
16. Hussein, I., Stipanović, D.: Effective coverage control for mobile sensor networks with guaranteed collision avoidance. IEEE Trans. Control Syst. Technol. **15**(4), 642–657 (2007)
17. Li, X., Jilkov, V.: Survey of maneuvering target tracking. Part I. Dynamic models. IEEE Trans. Aerosp. Electron. Syst. **39**(4), 1333–1364 (2003)
18. Naidu, D.: Optimal Control Systems, 1st, edn, vol. 2, pp.152–166. CRC Press (2002)

Chapter 8
Coverage-Based Cooperative Routing Algorithm for Unmanned Ground Vehicles

8.1 Introduction

The ability of traffic networks to support an increasing amount of vehicular traffic is becoming crucially important, as there are social, environmental and economic consequences of poorly managed networks [1]. In particular, vehicular congestion significantly jeopardizes the performance of traffic networks. A recent analysis of road speeds in New York suggests that it suffers from a rush hour which lasts all day [2]. There are a number of factors that may contribute to congestion on a traffic network. The most obvious is an increase in the number of vehicles using the network exceeding its capacity, for example during peak periods [3]. If some roads become congested, congestion will tend to spread to other parts of the network, so the faster those roads can be cleared, the better the network performs.

The advent of automated vehicles offers unique opportunities to reduce congestion, as control of the route is removed from human hands. Improving vehicle routes, according to some objective function, could become a new and powerful control action that can be taken. The dynamic vehicle routing problem has been concerned with this for some time [4, 5], but there are few clearly defined decentralised and adaptive control strategies for vehicles.

In a vehicular network, the routing problem is to find strategies to assign routes to vehicles in order to minimise their travel time and reduce congestion on the network. Strategies can be local, if they only rely on information available to the vehicle in a local neighbourhood of the network, or global if vehicles are assumed to know the state of the entire network. The simplest and most obvious global strategy is Dijkstra's algorithm [6], whereby each vehicle takes the shortest path between origin and destination, and its extensions (see for example [7, 8]). All of these modern routing algorithms enable extremely fast calculation of shortest paths (down to milliseconds), but all of them require some pre-processing of the traffic network and storage of data that is produced from this pre-processing. They are excellent for large static networks, but when a network is dynamic, such approaches are unpractical for

C. Zhai et al., *Cooperative Coverage Control of Multi-Agent Systems and its Applications*, Studies in Systems, Decision and Control 408,
https://doi.org/10.1007/978-981-16-7625-3_8

large networks, and require a high communications overhead when computed by a central controller and communicated to every vehicle.

Developments have been made for truly novel routing strategies. For instance in [9, 10], algorithms were developed for vehicle navigation, which aim to provide highly reliable travel time estimations. The algorithm, named 'Hyperstar', plans routes based on the probability of encountering a delay and needing to divert along a different path. In [11], a distributed algorithm is presented which can calculate the socially optimum set of traffic flows, given a set of vehicle trips, and then route vehicles probably using those optimum flows.

The aim of this chapter is to introduce and explore a simpler, yet effective, local routing algorithm based on using not only shortest path information but also some knowledge of localised congestion on different roads. The idea is that as a vehicle approaches a junction, it is assigned the next road on its route by evaluating a cost function composed of two terms: one related to distance from their destination, the other on the measured congestion level of roads out of that junction. The approach is inspired by coverage control strategies in control theory [12, 13] and is compared to other routing strategies on a set of representative examples. We find that, when compared to more traditional strategies, our approach guarantees a lower level of congestion and minimal delays. In particular, we uncover a subtle relationship between the mean travel time, the level of congestion and the weight in the cost function determining the relative dominance of the distance term over the congestion term. This seems to suggest that optimal tuning of routing control parameters is possible depending on the structural properties of the road network and the car density on it.

8.2 Basic Idea

Coverage control is a method for controlling distributed mobile sensor networks, in order that they can deploy mobile sensors and provide optimal coverage of a sensory function within a bounded region [14]. In order to maximise the flow in a traffic network, we hypothesize that it would be preferable to distribute vehicles as evenly as possible along roads, in order to reduce congestion while keeping their travel time minimal. This hypothesis is supported by the traffic flow model proposed by Lighthill and Whitham [15], which relates traffic density (the number of cars per km of road) to traffic flow (the number of cars passing a particular point per unit time), as well as by further approaches which have used congestion avoidance as a factor for calculating optimal routes for vehicles [16, 17]. In our routing algorithm, vehicles attempt to balance two sensory functions, one which attracts them towards their destination, and another which repels them from roads already heavily occupied by other vehicles (see Fig. 8.1).

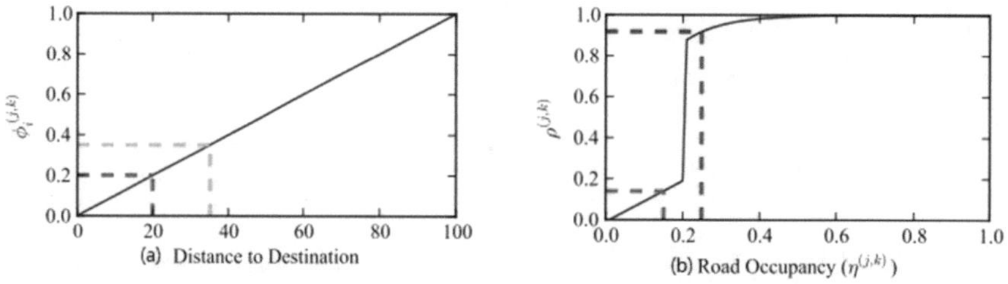

Fig. 8.1 Sensory functions based on distance and congestion levels

8.3 Routing Algorithm

Consider a set of N vehicles, $I = \{1, 2, ..., N\}$ in a bounded region $Q \subset R^2$. The position of the kth vehicle is denoted by $p_k \in Q$, $\forall k \in I$. The destination of the kth vehicle is denoted by some point $d_k \in Q$. All the vehicles are located on a road network $G = \{V, E\}$ with $V = \{1, 2, ..., m\}$ representing the set of m vertices (i.e., road junctions). E refers to the set of edges that stand for roads between junctions. Let v_i denote the ith junction, and $(v_i, v_j) \in E$ denotes the road between the junctions v_i and v_j. Each road (v_i, v_j) has three relevant properties:

- $L^{(i,j)}(t)$ quantifies the number of vehicles on the road (v_i, v_j) at time t
- $C^{(i,j)}$ denotes the maximum number of vehicles that can contain on the road (v_i, v_j)
- $\eta^{(v_i,v_j)}(t) = L^{(i,j)}(t)/C^{(i,j)}$ represents the percentage of space occupied by vehicles on the road (v_i, v_j) at time t

For example, if the kth vehicle arrives at the junction v_i, it can obtain the information on distance to the destination and congestion level, which are described by two sensory functions $\phi_k^{(i,j)}$ and $\rho^{(i,j)}$, respectively. The first sensory function $\phi_k^{(i,j)}$ is related to the estimated distance that the kth vehicle will travel if it takes a particular road (v_i, v_j) on the way to its destination:

$$\phi_k^{(i,j)} = \phi(d_k, v_i, v_j), \quad \phi_k^{(i,j)} \in [0, 1] \tag{8.1}$$

The second function $\rho^{(i,j)}$ is related to the occupancy $\eta^{(i,j)}(t)$ of road (v_i, v_j) at time t:

$$\rho^{(i,j)} = \rho(\eta^{(i,j)}(t)), \quad \rho^{(i,j)} \in [0, 1] \tag{8.2}$$

Equations (8.1) and (8.2) are combined with a tuning parameter $\alpha \in [0, 1]$ to construct a cost function $J_k^{(i,j)}$. When the kth vehicle reaches the junction v_i, it computes the cost function for every road (v_i, v_j) as follows

$$J_k^{(i,j)} = \alpha\phi^{(i,j)} + (1 - \alpha)\rho^{(i,j)} \tag{8.3}$$

The kth vehicle will take the road (v_i, v_j) that can minimize $J_k^{(i,j)}$. In practice, the function $\phi_k^{(i,j)}$ is selected as follows

$$\phi_k^{(i,j)} = \frac{D(v_i, d_k) + D(v_i, v_j)}{\max_{l \in V} D(v_l, d_k) + \max_{(s,t) \in E} D(v_s, v_t)}$$

where $D(v_i, d_k)$ is the network distance or shortest path between junction v_i and the destination d_k. $D(v_i, v_j)$ is the distance of road (v_i, v_j). In addition, $\max_{l \in V} D(v_l, d_k)$ refers to the largest network distance between the arbitrary junction v_l and the destination d_k. In comparisons, $\max_{(s,t) \in E} D(v_s, v_t)$ denotes the largest distance of the road in the traffic network. In order to take into account the congestion level, the function $\rho^{(i,j)}$ is equal to $\eta^{(i,j)}(t)$ when $\eta^{(i,j)}(t) \leq \eta_c^{(i,j)}$, and it is equal to $1 - e^{-\sigma \eta^{(i,j)}(t)}$ when $\eta^{(i,j)}(t) > \eta_c^{(i,j)}$. Here, $\eta_c^{(i,j)}$ is the critical occupancy of road (v_i, v_j) and σ is a tuning parameter to determine the sensitivity of $\rho^{(i,j)}$. The choice of σ is intended to emphasise the strong cost of going above $\eta_c^{(i,j)}$, with an exponential decay used so that we maintain some distinction between the cost functions for a set of roads which have all gone above their critical occupancy. It is worth pointing out that the above sensory functions can be selected differently and be made dependent on diverse routing factors such as travel time, congestion level and road condition.

8.4 Simulation Results

In order to validate the strategy, we carried out numerical simulations on some representative examples, first in Matlab and then using Simulation of Urban Mobility (SUMO) [18], a widely recognised micro agent simulator for traffic networks. We compare the performance of the coverage based routing algorithm with others based only on shortest path computation. In particular, our algorithm is compared with Dijkstra's and a modified Dijkstra algorithm discussed in [19] in the context of com-

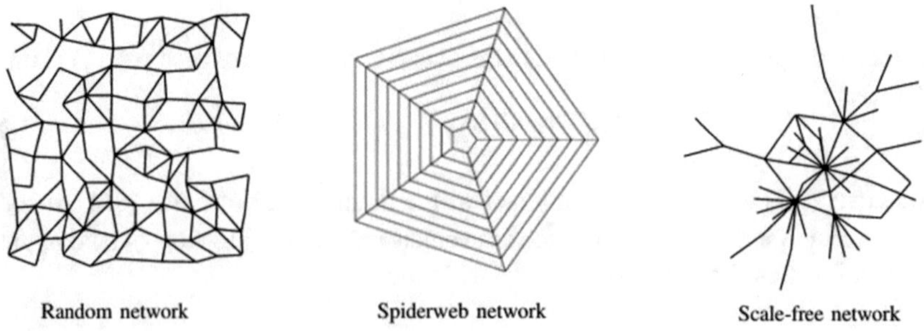

Fig. 8.2 Three different types of traffic networks

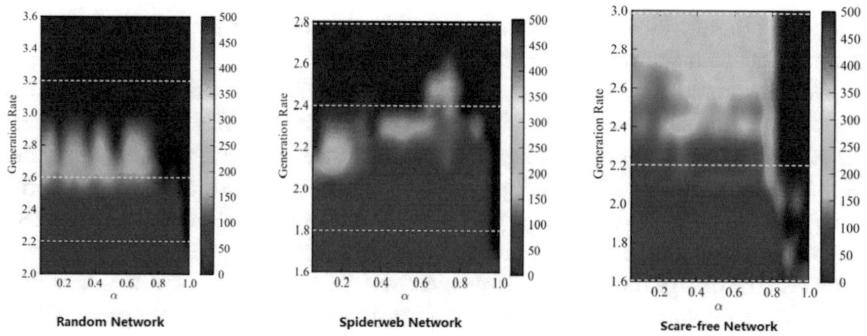

Fig. 8.3 Mean delay against α and the generation rate of new vehicles

Fig. 8.4 Cut-throughs of colormap figures

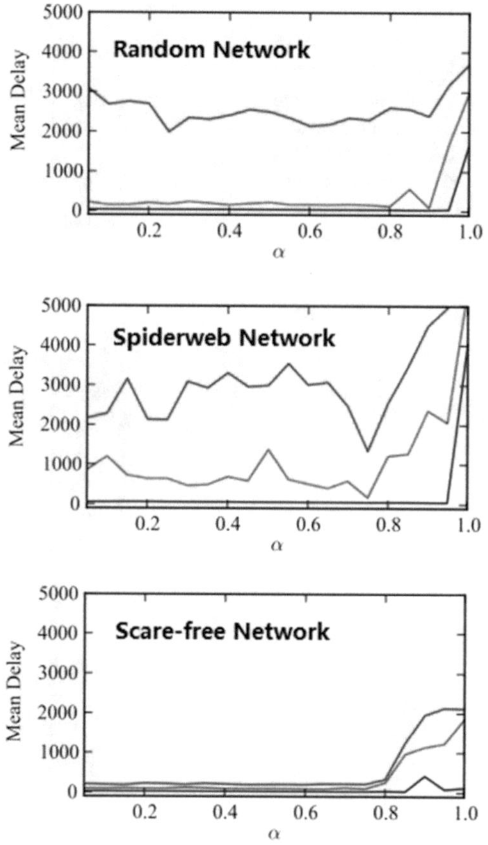

munication networks that, when more than one shortest path is available, takes the one with the next shortest queue of data packets.

We have compared our routing algorithm against routing using only a shortest path calculation, as calculated using Dijkstra's algorithm. We find that coverage based routing outperforms routing using only the shortest path, provided the value of α is tuned correctly. Figure 8.2 presents three different types of traffic networks, including random network, spiderweb network and scale-free network. Figure 8.3 shows the simulation results on mean delay against α and the generation rate of new vehicles. The orange indicates the large mean delays of vehicles in the corresponding network. In addition, Fig. 8.4 presents the Cut-throughs of colormap figures, which indicates that a larger value of α may lead to a high mean delay of travel time for the vehicles.

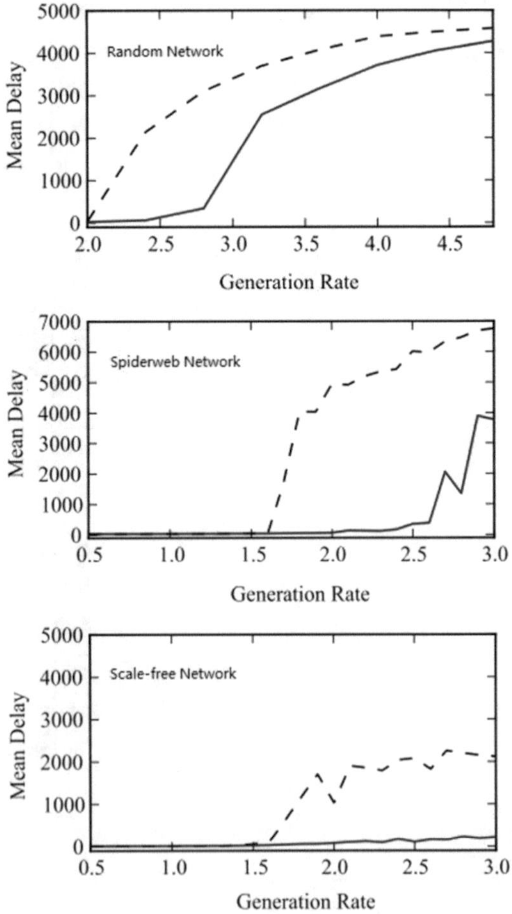

Fig. 8.5 Mean delay of vehicles on different traffic networks with respect to the generation rate of new vehicles. The dashed black lines refer to the routing using the shortest path, and the solid blue lines indicate the coverage based routing algorithm with an optimal value of α in Table III

Figure 8.5 show shortest path based routing and coverage based routing exhibiting similar delays at low car generation rates. However, at some intermediate car generation rates we see shortest path routing unable to keep the network free from congestion, and delays increase by an order of magnitude. In contrast, coverage based routing can continue to maintain much lower delays up to some other, higher, car generation rate. This demonstrates that coverage based routing increases the capacity of the network over routing using only the shortest path. In terms of the increase in maximum car generation rate that coverage based routing exhibits over shortest path routing, we see around a 50% increase for the 5×5 grid network, a 120% increase in the 10×10 grid network, a 40% increase in the random network, a 60% increase in the spiderweb network, and the absence of a definitive maximum car generation rate in the scale-free network (at least over the range of car generation rates being considered). The simulations indicate that significant gains in network capacity can be made by using a coverage based approach over shortest path routing, however the size of these gains is dependent upon network topology.

8.5 Conclusions

The proposed routing strategy shows how a simple cost function coupling global information about distances with local road occupancy data can yield improvements over shortest path routing. These improvements are an increase in the capacity of the network, and hence the ability to avoid congestion which leads to delays. We have also shown that in order to optimise this routing strategy a control parameter must be chosen appropriately, and this value appears to depend on the network topology. We envisage that the algorithm could be deployed via a junction infrastructure able to communicate to each vehicle arriving at the junction the best road to take next. We wish to emphasise that the algorithm we present here can be deployed effectively and is easily scalable because of its simplicity. Future work will address the possibility of each car, or junction, becoming able to tune the control parameter α in real time via local adaptive strategies aimed at further minimising congestion and guarantee fairness. Also, the routing strategy will be tested on real road network topologies, using realistic loading on the network, including a variety of vehicle types.

References

1. Graham, D.: Variable returns to agglomeration and the effect of road traffic congestion. J. Urban Econ. **62**, 103–120 (2007)
2. Wellington, B.: Quantifying the best and worst times of day to hit the road in NYC. Technical Report (2014)
3. Department for Transportation, Congestion: A National Issue. Technical Report (2014)
4. Pillac, V., Gendreau, M., Gueret, C., Medaglia, A.: A review of dynamic vehicle routing problems. Eur. J. Oper. Res. **225**, 1–11 (2013)

5. Smith, M.: The stability of a dynamic model of traffic assignment - an application of a method of Lyapunov. Transp. Sci. **18**, 245–252 (1984)
6. Dijkstra, E.: A note on two problems in connexion with graphs. Numer. Math. **1**, 269–271 (1959)
7. Hart, P., Nils, J.: Formal basis for the heuristic determination of minimum cost paths. IEEE Trans. Syst. Cybern. **4**(2), 100–107 (1968)
8. Delling, D., Sanders, P., Schultes, D., Wagner, D.: Engineering route planning algorithms. Algorithmics **2**, 117–139 (2009)
9. Bell, M.: Hyperstar: a multi-path Astar algorithm for risk averse vehicle navigation. Transp. Res. Part B: Methodol. **43**, 97–107 (2009)
10. Bell, M., Trozzi, V., Hosseinloo, S., Gentile, G., Fonzone, A.: Time-dependent hyperstar algorithm for robust vehicle navigation. Transp. Res. Part A: Policy Pract. **46**, 790–800 (2012)
11. Lim, S., Rus, D.: Stochastic distributed multi-agent planning and applications to traffic. In: Proceedings of IEEE International Conference on Robotics and Automation, pp. 2873–2879 (2012)
12. Schwager, M., Rus, D., Slotine, J.: Decentralized, adaptive coverage control for networked robots. Int. J. Robot. Res. **28**(3), 357–375 (2009)
13. Zhai, C., He, F., Hong, Y., Wang, L., Yao, Y.: Coverage-based interception algorithm of multiple interceptors against the target involving decoys. J. Guid. Control. Dyn. **39**(7), 1647–1653 (2016)
14. Schwager, M., Mclurkin, J., Rus, D.: Distributed coverage control with sensory feedback for networked robots. In: Proceedings of Robotics: Science and Systems, pp. 49–56 (2006)
15. Lighthill, M., Whitham, G.: On kinematic waves II. A theory of traffic flow on long crowded roads. Proc. R. Soc. Lond. Ser. A, Math. Phys. Sci. **229**(1178), 317–345 (1955)
16. Smitha, S., Narendra, K., Usha, R., Divyashree, C., Gayatri, G., Aparajitha, M.: GPS based shortest path for ambulances using VANETs. In: 2012 International Conference on Wireless Networks, vol. 49, pp. 190–196 (2012)
17. Aslam, J., Lim, S., Rus, D.: Congestion-aware traffic routing system using sensor data. In: 15th International IEEE Conference on Intelligent Transportation Systems, pp. 1006–1013 (2012)
18. Behrisch, M., Bieker, L., Erdmann, J., Krajzewicz, D.: Sumo (simulation of urban mobility). In: The Third International Conference on Advances in System Simulation, pp. 55–60 (2011)
19. Manfredi, S., Garofalo, F., Bernardo, M.: Analysis and effects of retransmission mechanisms on data network performance. In: Proceedings of the 2004 International Symposium on Circuits and Systems, pp. 625–628 (2004)

Chapter 9
Cooperative Coverage Control of Wireless Sensor Networks for Environment Monitoring

9.1 Introduction

In the past decades, multi-agent systems and cooperative control have received increasingly attention in a wide range of applications, which play an important role in distributed control and information collection. Multi-agent systems provide an useful framework to investigate the collective phenomenon in both nature and engineering such as flocking, consensus, and coverage [1–3].

Cooperative coverage of multiple agents in the environment derives from many practical problems such as exploration, monitoring, surveillance and environmental monitor [4–6], which has drawn much attention to the researchers in various fields. Up to now, different optimization formulations have been proposed for coverage problem [7–12]. For example, [9] considered the sweep coverage in the uncertain environment and provided a decentralized coverage algorithm, which incorporates two operations: workload partition and sweeping. Reference [10] designed the ant robot, which uses smell traces to mark the covered areas and help navigation. By this method, complete coverage of the region can be achieved even if the environment changes. Reference [11] described a multi-robot coverage algorithm based on cellular decomposition, and two types of robots are used to cover the free space and determine the termination of the cell. In addition, the coverage algorithm of sensor networks was designed to monitor random events with a frequency density function and maximize the joint detection probabilities of random events in [12].

As a special type of potential functions, navigation functions have been widely used to deal with motion planning problems in robotics. For example, [13] constructed a class of navigation functions on analytic manifolds with boundary, which are composed of the goal function, the constraint function and a tuning parameter, and proved that the trajectory of system converges to the minima of the goal function with constraints when the tuning parameter is sufficiently large. Reference [14] presented a communication-aware navigation framework of multiple robot system, which is used to track a moving target while avoiding collision with both fixed and mobile

C. Zhai et al., *Cooperative Coverage Control of Multi-Agent Systems and its Applications*, Studies in Systems, Decision and Control 408,
https://doi.org/10.1007/978-981-16-7625-3_9

obstacles. Moreover, [15] proposed a decentralized cooperative controller for multiple agents to generate a formation in the given region while avoiding obstacles and collisions. Furthermore, [16] considered formation control of multiple agents while maintaining the global network connectivity and avoiding obstacles by designing the decentralized navigation function. The main contribution of this work is two-fold. On the one hand, we propose a new coverage formulation of multiple agents in the environment. On the other hand, we extend the application of navigation functions by investigating the coverage of multiple agents in the region with obstacles. This extension results in the undesired critical points of the navigation function, and we prevent the trajectory of the system from converging to these points by designing the constraint function.

The rest of this chapter is organized as follows. In Sect. 9.2, a coverage formulation of multiple agents in the given region is presented. In Sect. 9.3, the theoretical analysis is conducted. Firstly, coverage control of agents in the region without obstacles is discussed. Then, with the help of navigation functions, a decentralized coverage algorithm is also proposed to handle coverage problem in the region with obstacles. Numerical simulations are carried out in Sect. 9.4. Finally, we draw a conclusion in the last section.

9.2 Problem Statement

In this section we propose our coverage formulation. Consider n point-mass agents equipped with sensors that detect the random event taking place in the mission space $\mathscr{W} \in R^2$, and $\mathscr{W}$ is a bounded, closed and convex region. $\phi(q)$ is a continuous event density function (EDF) or probability distribution function (PDF), which captures the frequency of a random event taking place at the position q. $\phi(q) \geq 0, \forall q \in \mathscr{W}$ ($\phi(q) = 0, \forall q \notin \mathscr{W}$), and it satisfies

$$\int_{\mathscr{W}} \phi(q)dq < \infty.$$

$s_i \in R^2$ denotes the position of agent $i \in E_n = \{1, 2, ..., n\}$. Once a random event appears at q, it will be detected by agent i if $\|s_i - q\| \leq r_s$. Here, r_s is the sensing radius and $\|s_i - q\|$ represents the distance between the random event and agent i. Then, agent i will take the corresponding measures to deal with the random event. The intensity of the action imposed by agent i (i.e., $p(\|s_i - q\|)$) decreases with the distance $\|s_i - q\|$, and its effectiveness is denoted by $\mu(p(\|s_i - q\|))$. Hence, the joint effectiveness of actions imposed by multiple agents for the random event occurring at q can be written as

$$f(s, q) = \mu\left(\sum_{i=1}^{n} p(\|s_i - q\|)\right)$$

where $s = (s_1^T, s_2^T, ..., s_n^T)^T$. Therefore, the expected effectiveness for the random event by n mobile agents deploying in the mission space $\mathscr{W}$ is expressed as

$$H(s) = \int_{\mathscr{W}} \phi(q) f(s, q) dq$$

Remark 9.1 The function $p : R^+ \to [0, 1]$ is continuously differentiable and satisfies

1. $p(0) = 1$;
2. $p(x) = 0, \forall x \geq r_s$;
3. $p'(x) \leq 0, x \in [0, r_s]$.

where $p'(x) = \frac{dp(x)}{dx}$. Similarly, the continuously differentiable function $\mu : R^+ \to [0, 1]$ has the following properties:

1. $\mu(0) = 0$;
2. $\mu(x) = 1, \forall x \geq 1$;
3. $\mu'(x) \geq 0, x \in [0, 1]$.

Intuitively, if the random event is regarded as the fire in the forest, and each agent will stand for a fire station, where the fire airplane takes off and goes to put out a fire. $p(x)$ denotes the largest amount of water loaded on the airplane, and it is related to the flight distance x. Moreover, $\mu(x)$ shows the actual effect of the amount of water on the fire.

Notice that each agent only has local sensing capability with the sensing radius r_s. The coverage region of agent i is denoted by

$$\Omega_i = \{q \in R^2 : \|q - s_i\| \leq r_s\}.$$

Thus, the neighbor set of agent i is defined as

$$N_i = \{j \in E_n : \|s_j - s_i\| \leq 2r_s\}.$$

The agent cannot deal with the random event taking place outside of its coverage region (i.e., $p(\|s_i - q\|) = 0, \forall q \notin \Omega_i$). In addition, we have $p(\|s_j - q\|) = 0, \forall q \in \Omega_i$ if agent j is not the neighbor of agent i. Hence, the gradient of $H(s)$ with respect to s_i can be computed as

$$\begin{aligned}
&\nabla_{s_i} H \\
&= \int_{\mathscr{W}} \phi(q) \nabla_{s_i} f(s, q) dq \\
&= \int_{\mathscr{W}} \phi(q) \mu' \left(\sum_{j=1}^{n} p(\|s_j - q\|) \right) p'(\|s_i - q\|) \frac{s_i - q}{\|s_i - q\|} dq \\
&= \int_{\Omega_i} \phi(q) \mu' \left(\sum_{j \in N_i} p(\|s_j - q\|) \right) p'(\|s_i - q\|) \frac{s_i - q}{\|s_i - q\|} dq
\end{aligned}$$

Since both $p(x)$ and $\mu(x)$ are continuously differentiable, $\nabla_{s_i} H$ exists everywhere. In this paper, our objective is to develop a decentralized control law that fixes the position of each agent in the given region in order to maximize $H(s)$.

9.3 Main Results

In this section, we discuss the coverage behaviors of multiple agents in two kinds of coverage regions.

9.3.1 *Coverage Region Without Obstacles*

At first, we give the theoretical result on the coverage in the region without obstacles.

Theorem 9.1 *For each agent with the following dynamics*

$$\dot{s}_i = \nabla_{s_i} H, \quad i \in E_n,$$

we have

1. $\lim_{t \to +\infty} \nabla_{s_i} H = 0, i \in E_n$;
2. *The isolated maximum of $H(s)$ is the asymptotically stable equilibrium of the dynamical system.*

Proof Take $V(s) = \max_{s \in \mathscr{W}^n} H(s) - H(s) \geq 0$, and thus, the time derivative of $V(s)$ along the trajectory of the system is given by

$$\dot{V}(s) = -\dot{H}(s) = -\sum_{i=1}^{n} \nabla_{s_i} H^T \dot{s}_i = -\sum_{i=1}^{n} \|\nabla_{s_i} H\|^2 \leq 0$$

Obviously, $\mathscr{W}^n$ is a compact set since $\mathscr{W}$ is bounded and closed. Moreover, $\mathscr{W}^n$ is positively invariant with respect to the dynamical system because when $s_i \in \partial \mathscr{W}$, we have

$$\nabla_{s_i} H = \int_{\Omega_i \cap \mathscr{W}} \phi(q) \mu' \left(\sum_{j \in N_i} p(\|s_j - q\|) \right) p'(\|s_i - q\|) \frac{s_i - q}{\|s_i - q\|} dq$$

Because $\phi(q) \geq 0$ and $\mu'(\sum_{j \in N_i} p(\|s_j - q\|)) \geq 0$ as well as $p'(\|s_i - q\|) \leq 0$, we get $\nabla_{s_i} H = 0$ or $\nabla_{s_i} H$ is directed to the interior of the region $\mathscr{W}$. It means that agent i stays on the boundary of $\mathscr{W}$ or it moves towards the interior of $\mathscr{W}$. With the aid of LaSalle's invariance principle, we conclude that the trajectory of the system converges to the largest invariant set

$$S = \{s^* \in \mathscr{W}^n : \dot{V}(s^*) = 0\} = \{s^* \in \mathscr{W}^n : \nabla_{s_i} H|s^* = 0, i \in E_n\}.$$

For the isolated maximum $\tilde{s}$, there exists a neighborhood $U(\tilde{s})$ of $\tilde{s}$ such that $H(s) < H(\tilde{s})$ and

$$\nabla_{s_i} H \neq 0, i \in E_n, \forall s \in U(\tilde{s}) - \{\tilde{s}\}.$$

Choose

$$\widetilde{V}(s) = H(\tilde{s}) - H(s), s \in U(\tilde{s})$$

as the Lyapunov function candidate, then we get $\widetilde{V}(s) > 0$ and $\dot{\widetilde{V}}(s) < 0$, $s \in U(\tilde{s}) - \{\tilde{s}\}$. Hence, $\tilde{s}$ is asymptotically stable. The proof of this theorem is thus completed. □

Remark 9.2 The theoretical formulation is somewhat similar to some previous work on coverage like [3, 8, 12]. Nevertheless, our formulation derives from distinct backgrounds and focuses on the implementation of multiple agent systems rather than the sensing capability.

Remark 9.3 The convexity of the region $\mathscr{W}$ is required to guarantee the positive invariance of $\mathscr{W}$ with respect to the dynamical system. For the non-convex region $\mathscr{W}$, we can choose $conv(\mathscr{W}^n)$ (i.e., the convex hull of $\mathscr{W}^n$) as the compact, positively invariant set. In addition, the saddle points and the minimum of $H(s)$ are unstable for the dynamical system.

Clearly, when $\nabla_s H|s^* = 0$ and $\nabla_s^2 H|s^* \prec 0$, $H(s)$ achieves its maximal value $H(s^*)$, where $s^* = (s_1^{*T}, s_2^{*T}, ..., s_n^{*T})^T$ is the critical point of $H(s)$ and the symbol "$\prec 0$" denotes the negative definiteness of the relevant matrix. Thus, we propose the assumption as follows: At the critical points of $H(s)$, we have $\nabla_s^2 H \prec 0$ or $\nabla_s^2 H \succ 0$. This assumption implies that the critical points of $H(s)$ correspond to the local maximums or minimums of $H(s)$.

Lemma 9.1 *The following two claims hold*

1. $\nabla_s H|s^* = 0 \Leftrightarrow \nabla_{s_i} H|s^* = 0, \forall i \in E_n$;
2. $\nabla_s^2 H|s^* \prec 0 \Rightarrow \nabla_{s_i}^2 H|s^* \prec 0, \forall i \in E_n$.

Proof From $\nabla_s H = (\nabla_{s_1} H^T, \nabla_{s_2} H^T, ..., \nabla_{s_n} H^T)^T$, the first claim holds naturally. Moreover, since

$$\nabla_s^2 H|s^* \prec 0 \Longleftrightarrow -\nabla_s^2 H|s^* \succ 0,$$

we have $(-\nabla_s^2 H|s^*)_{ii} > 0$ and $\det(-\nabla_{s_i}^2 H|s^*) > 0, \forall i \in E_n$ because all principle minors of a positive definite matrix are positive. Thus, we obtain $(\nabla_s^2 H|s^*)_{ii} < 0$ and $\det(\nabla_{s_i}^2 H|s^*) > 0$. It follows from

$$\nabla_{s_i}^2 H|s^* = \begin{pmatrix} (\nabla_s^2 H|s^*)_{2i-1,2i-1} & (\nabla_s^2 H|s^*)_{2i-1,2i} \\ (\nabla_s^2 H|s^*)_{2i,2i-1} & (\nabla_s^2 H|s^*)_{2i,2i} \end{pmatrix}$$

that $\nabla_{s_i}^2 H|s^* \prec 0, \forall i \in E_n$. □

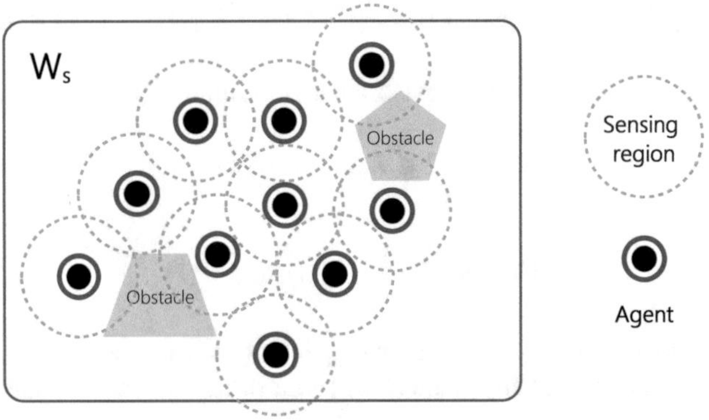

Fig. 9.1 Coverage of multiple agents in the sphere region $\mathscr{W}_s$ with obstacles

9.3.2 Coverage Region with Obstacles

Let us consider the coverage problem in the region with m obstacles. For simplicity, the sphere region $\mathscr{W}_s = \{q \in R^2 : \|q\| \leq R\}$ with the radius R and the center located at the origin is considered in the sequel (see Fig. 9.1). Actually, other regions that can be transformed to $\mathscr{W}_s$ through diffeomorphism are also feasible, because diffeomorphisms preserve the properties of navigation functions [17]. Let q_j denote the center of the sphere obstacle j with the radius r_j, and the obstacle j is expressed as $O_j = \{q \in \mathscr{W}_s : \|q - q_j\| < r_j\}$, $j \in E_m$. Moreover, the obstacles satisfy $\|q_i - q_j\| > r_i + r_j, \forall i, j \in E_m$, which means none of them intersect. Further, all obstacles are contained in the interior of the sphere region (i.e., $\|q_j\| + r_j < R, \forall j \in E_m$). Then we introduce the decentralized navigation function[1] candidate as follows [13]

$$\varphi_i = \frac{\gamma_i}{(\gamma_i^\alpha + \beta_i)^{\frac{1}{\alpha}}}$$

where $\alpha > 0$ is a tuning parameter, $\gamma_i = \|\nabla_{s_i} H\|^2$ is the goal function for agent i, and the constraint function β_i is given as follows:

$$\beta_i = \beta_{ib} \cdot \beta_{iu}^{\kappa} \cdot \prod_{j \in M_i} \beta_{ij}$$

where $\beta_{iu}^{\kappa} = (\beta_{iu})^{\kappa}$ with a positive tuning parameter κ,

$$M_i = \{j \in E_m : 0 \leq \|s_i - q_j\| - r_j \leq \delta_1\}$$

[1] Let $\mathscr{D}$ be a compact, connected analytic manifold with boundary. A map $\varphi : \mathscr{D} \to [0, 1]$ is a navigation function if it is analytic, polar, Morse and admissible on $\mathscr{D}$ [13].

and β_{ij} prevents agent i from colliding with obstacle $j \in E_m$ constructed as:

$$\beta_{ij} = \begin{cases} \sin(\frac{\pi}{2\delta_1}(\|s_i - q_j\| - r_j)), & \|s_i - q_j\| - r_j \leq \delta_1; \\ 1, & \|s_i - q_j\| - r_j > \delta_1. \end{cases}$$

Likewise,

$$\beta_{ib} = \begin{cases} \sin(\frac{\pi}{2\delta_1}(R - \|s_i\|)), & R - \|s_i\| \leq \delta_1; \\ 1, & R - \|s_i\| > \delta_1. \end{cases}$$

is used to avoid a collision with the boundary of the mission space. Moreover, β_{iu} is designed as

$$\beta_{iu} = g\left(\frac{\partial^2 H}{\partial s_{ix}^2}\right) + g(-\|\nabla_{s_i} H\|),$$

where $s_i = (s_{ix}, s_{iy})^T$ and

$$g(x) = \begin{cases} c(1 - e^{-x^2}), & x \leq 0; \\ 0, & x > 0. \end{cases}$$

with the tuning parameter $c > 0$.

Remark 9.4 The navigation function φ_i gets its global maximum 1 when agent i arrives at undesired positions or collides with the boundary of the mission space or obstacles (i.e., $\beta_i = 0$). In addition, φ_i achieve its global minimum 0 if $\gamma_i = 0$.

Remark 9.5 The positive scalar δ_1 with $\delta_1 + r_j \leq 2r_s$, $j \in E_m$ denotes the repulsive range between the agent and its adjacent obstacles, and it can be adjusted as small as possible.

Remark 9.6 β_{iu} drives the trajectory of the system away from undesired critical points of $H(s)$.

Furthermore, each agent takes the following dynamics

$$\dot{s}_i = u_i, \quad i \in E_n \tag{9.1}$$

where the control law u_i is of the form

$$u_i = -K\nabla_{s_i}\varphi_i \tag{9.2}$$

with the scalar gain $K > 0$. To facilitate the theoretical analysis, the free workspace

$$\mathscr{W}_i = \left\{ s : \tilde{\beta}_i(s) > 0, s_j \in \mathscr{W}_s - \bigcup_{i=1}^{m} O_i, j \in E_n \right\}$$

for agent i is partitioned into the following subsets, where $\tilde{\beta}_i = \beta_{ib} \cdot \prod_{j \in M_i} \beta_{ij}$. The boundary of $\mathscr{W}_i$ is given by $\partial \mathscr{W}_i = \tilde{\beta}_i^{-1}(0)$. The set of destination points for agent i is described as

$$F_{id} = \{s : \nabla_{s_i} H|s = 0, \beta_{iu}(s) \neq 0\}.$$

Let

$$B_{ib}(\epsilon) = \{s : 0 < \beta_{ib} < \epsilon\}, \quad B_{iu}(\epsilon) = \{s : 0 < \beta_{iu}^{\kappa} < \epsilon\}$$

and

$$B_{ij}(\epsilon) = \{s : 0 < \beta_{ij} < \epsilon, j \in M_i\}.$$

Then $F_0(\epsilon) = B_{ib}(\epsilon) - F_{id}$ denotes the region near the boundary of workspace, and $F_1(\epsilon) = \bigcup_{j \in M_i} B_{ij}(\epsilon) - F_{id}$ represents the region around the potential collision with adjacent obstacles. In addition, $F_2(\epsilon) = B_{iu}(\epsilon) - F_{id}$ describes the region near the undesired critical points. Clearly,

$$F_3(\epsilon) = \mathscr{W}_i - \left\{F_{id} \bigcup F_0(\epsilon) \bigcup F_1(\epsilon) \bigcup F_2(\epsilon)\right\}$$

is the complement of the above subsets. The gradient of φ_i with respect to s_i is calculated as follows:

$$\nabla_{s_i} \varphi_i = \frac{\alpha \beta_i \nabla_{s_i} \gamma_i - \gamma_i \nabla_{s_i} \beta_i}{\alpha (\gamma_i^{\alpha} + \beta_i)^{\frac{1}{\alpha}+1}} \tag{9.3}$$

To facilitate the analysis, we have the following assumption

$$\frac{\gamma_i}{\|\nabla_{s_i} \gamma_i\|} < \infty, \quad \forall s \in F_3(\epsilon).$$

Geometrically, this assumption indicates that, for every point in $F_3(\epsilon)$, the gradient of $H(s)$ with respect to s_i are not perpendicular to the plane which is generated by the two row vectors of the Hessian matrix $\nabla_{s_i}^2 H$.

Lemma 9.2 *When $\kappa > \alpha$, φ_i achieves its maximum* 1 *at the undesired point*

$$s \in \left\{s : \nabla_{s_i} H|s = 0, \frac{\partial^2 H}{\partial s_{ix}^2}|_s > 0\right\}.$$

Proof φ_i can be rewritten as

$$\varphi_i = \frac{1}{(1 + \frac{\beta_i}{\gamma_i^{\alpha}})^{\frac{1}{\alpha}}} = \frac{1}{(1 + \beta_{ib} \cdot \prod_{j \in M_i} \beta_{ij} \cdot \frac{\beta_{iu}^{\kappa}}{\gamma_i^{\alpha}})^{\frac{1}{\alpha}}}$$

where

$$\frac{\beta_{iu}^{\kappa}}{\gamma_i^{\alpha}} = \frac{[g(\frac{\partial^2 H}{\partial s_{ix}^2}) + g(-\|\nabla_{s_i} H\|)]^{\kappa}}{\gamma_i^{\alpha}}$$
$$= \frac{[g(\frac{\partial^2 H}{\partial s_{ix}^2}) + c(1 - e^{-\gamma_i})]^{\kappa}}{\gamma_i^{\alpha}}$$

Since $\frac{\partial^2 H}{\partial s_{ix}^2} > 0$ at the undesired points and it is continuous on the free workspace, there is a scalar $\delta_2 > 0$ such that the sign of $\frac{\partial^2 H}{\partial s_{ix}^2}$ keeps unchanged if $\gamma_i < \delta_2$. Therefore,

$$\lim_{\gamma_i \to 0} \frac{\beta_{iu}^{\kappa}}{\gamma_i^{\alpha}} = \lim_{\gamma_i \to 0} \frac{c^{\kappa}(1 - e^{-\gamma_i})^{\kappa}}{\gamma_i^{\alpha}} = 0$$

and thus φ_i tends to 1 when $\kappa > \alpha$. □

Lemma 9.3 *For every $\epsilon > 0$, there exists $\eta(\epsilon)$ such that there are no critical points of φ_i in $F_3(\epsilon)$ if $\alpha > \eta(\epsilon)$.*

Proof When $\nabla_{s_i} \varphi_i = 0$, we have $\alpha \beta_i \nabla_{s_i} \gamma_i = \gamma_i \nabla_{s_i} \beta_i$ according to (9.3). If α satisfies the following inequality

$$\alpha > \sup_{s \in F_3(\epsilon)} \frac{\gamma_i \|\nabla_{s_i} \beta_i\|}{\|\nabla_{s_i} \gamma_i\| \beta_i},$$

the equality $\alpha \beta_i \nabla_{s_i} \gamma_i = \gamma_i \nabla_{s_i} \beta_i$ does not hold in $F_3(\epsilon)$ any more. Therefore, there are no critical points of φ_i in $F_3(\epsilon)$.

$$\frac{\gamma_i \|\nabla_{s_i} \beta_i\|}{\|\nabla_{s_i} \gamma_i\| \beta_i} = \frac{\gamma_i}{\|\nabla_{s_i} \gamma_i\|} \cdot \frac{\|\nabla_{s_i} (\beta_{ib} \cdot \beta_{iu}^{\kappa} \cdot \prod_{j \in M_i} \beta_{ij})\|}{\beta_{ib} \cdot \beta_{iu}^{\kappa} \cdot \prod_{j \in M_i} \beta_{ij}}$$
$$\leq \frac{\gamma_i}{\|\nabla_{s_i} \gamma_i\|} \left(\frac{\|\nabla_{s_i} \beta_{ib}\|}{\beta_{ib}} + \frac{\|\nabla_{s_i} \beta_{iu}^{\kappa}\|}{\beta_{iu}^{\kappa}} + \frac{\|\nabla_{s_i} \prod_{j \in M_i} \beta_{ij}\|}{\prod_{j \in M_i} \beta_{ij}} \right)$$

Since $\beta_{ib} \geq \epsilon$, $\beta_{iu}^{\kappa} \geq \epsilon$ and $\beta_{ij} \geq \epsilon$ in $F_3(\epsilon)$, we get

$$\frac{\gamma_i \|\nabla_{s_i} \beta_i\|}{\|\nabla_{s_i} \gamma_i\| \beta_i} \leq \frac{\gamma_i}{\|\nabla_{s_i} \gamma_i\|} \left(\frac{\|\nabla_{s_i} \beta_{ib}\|}{\epsilon} + \frac{\|\nabla_{s_i} \beta_{iu}^{\kappa}\|}{\epsilon} + \frac{\|\nabla_{s_i} \prod_{j \in M_i} \beta_{ij}\|}{\epsilon^m} \right)$$

Thus, $\eta(\epsilon)$ is selected as follows:

$$\begin{aligned}\sup_{s\in F_3(\epsilon)} \frac{\gamma_i \|\nabla_{s_i}\beta_i\|}{\|\nabla_{s_i}\gamma_i\|\beta_i} \le & \sup_{s\in F_3(\epsilon)} \frac{\gamma_i}{\|\nabla_{s_i}\gamma_i\|}\left(\frac{\sup_{s\in F_3(\epsilon)}\|\nabla_{s_i}\beta_{ib}\|}{\epsilon}\right.\\ & + \frac{\sup_{s\in F_3(\epsilon)}\|\nabla_{s_i}\beta_{iu}^{\kappa}\|}{\epsilon}\\ & \left.+ \frac{\sup_{s\in F_3(\epsilon)}\|\nabla_{s_i}\prod_{j\in M_i}\beta_{ij}\|}{\epsilon^m}\right)\\ = & \eta(\epsilon)\end{aligned}$$

Hence, we complete the proof. □

Lemma 9.4 *There are no critical points of φ_i on $\partial\mathscr{W}_i$ if $\beta_{iu}(s) \neq 0, \forall s \in \partial\mathscr{W}_i$.*

Proof For any $\tilde{s}$ with $\tilde{s} \in \partial\mathscr{W}_i$, we have $\tilde{\beta}_i(\tilde{s}) = 0$. Moreover, $\nabla_{s_i}\varphi_i$ evaluated at $\tilde{s}$ is expressed as

$$\begin{aligned}\nabla_{s_i}\varphi_i|_{\tilde{s}} &= -\frac{1}{\alpha\gamma_i^{\alpha}}\nabla_{s_i}\beta_i|_{\tilde{s}}\\ &= -\frac{1}{\alpha\gamma_i^{\alpha}}\nabla_{s_i}\left(\beta_{ib}\cdot\beta_{iu}^{\kappa}\cdot\prod_{j\in M_i}\beta_{ij}\right)|_{\tilde{s}}\\ &= -\frac{1}{\alpha\gamma_i^{\alpha}}\left[\left(\nabla_{s_i}\beta_{ib}\cdot\beta_{iu}^{\kappa}\cdot\prod_{j\in M_i}\beta_{ij}\right)|_{\tilde{s}}\right.\\ &\quad + \left(\beta_{ib}\cdot\nabla_{s_i}\beta_{iu}^{\kappa}\cdot\prod_{j\in M_i}\beta_{ij}\right)|_{\tilde{s}}\\ &\quad \left.+ \left(\beta_{ib}\cdot\beta_{iu}^{\kappa}\cdot\nabla_{s_i}\prod_{j\in M_i}\beta_{ij}\right)|_{\tilde{s}}\right]\end{aligned}$$

Considering that none of the obstacles intersect and they are contained in the interior of $\mathscr{W}$, we have $\beta_{ib}(\tilde{s}) \neq 0$ if $\beta_{ij}(\tilde{s}) = 0, \forall j \in E_m$ and vice versa. Similarly, $\beta_{ij}(\tilde{s}) \neq 0$ if $\beta_{ik}(\tilde{s}) = 0, \forall j, k \in E_m$ and $j \neq k$. When $\beta_{ib}(\tilde{s}) = 0$, $\|\tilde{s}_i\| = R$ with $\tilde{s} = (\tilde{s}_1, \tilde{s}_2, ..., \tilde{s}_n)^T$. Thus,

$$\begin{aligned}\nabla_{s_i}\beta_{ib}|_{\tilde{s}} &= -\frac{\pi}{2\delta_1}\cos\left(\frac{\pi}{2\delta_1}(R - \|s_i\|)\right)\frac{s_i}{\|s_i\|}|_{\tilde{s}}\\ &= -\frac{\pi}{2\delta_1 R}\tilde{s}_i \neq 0\end{aligned}$$

Because $\beta_{iu}(s) \neq 0, \forall s \in \partial\mathscr{W}$, we obtain

$$\nabla_{s_i}\varphi_i|_{\tilde{s}} = -\frac{1}{\alpha\gamma_i^{\alpha}}\left(\nabla_{s_i}\beta_{ib}\cdot\beta_{iu}^{\kappa}\cdot\prod_{j\in M_i}\beta_{ij}\right)|_{\tilde{s}} \neq 0$$

Likewise, it follows from $\beta_{ij}(\tilde{s}) = 0, \forall j \in M_i$ that

$$\nabla_{s_i}\beta_{ij}|_{\tilde{s}} = \frac{\pi}{2\delta_1}\cos\left(\frac{\pi}{2\delta_1}(\|s_i - q_j\| - r_j)\right)\frac{s_i - q_j}{\|s_i - q_j\|}|_{\tilde{s}}$$
$$= \frac{\pi}{2\delta_1 r_j}(\tilde{s}_i - q_j) \neq 0$$

since $\|\tilde{s}_i - q_j\| = r_j$. We also have

$$\nabla_{s_i}\varphi_i|_{\tilde{s}} = -\frac{1}{\alpha\gamma_i^{\alpha}}\left(\beta_{ib}\cdot\beta_{iu}^{\kappa}\cdot\nabla_{s_i}\prod_{j\in M_i}\beta_{ij}\right)|_{\tilde{s}} \neq 0$$

Hence, there are no critical points of φ_i on $\partial\mathscr{W}_i$. □

Lemma 9.5 *There exists $\epsilon_1 > 0$ such that φ_i has no local minimum in the region*

$$F_0(\epsilon)\bigcup F_1(\epsilon)\bigcup F_2(\epsilon)$$

and all the critical points of φ_i in the above region are non-degenerate if $\epsilon < \epsilon_1$.

The proof of this lemma is similar to that of Proposition 3.6 and 3.9 in [13], and thus is omitted.

Lemma 9.6 *φ_i is minimized at the destination points $s \in F_{id}$.*

Proof Considering $\gamma_i = 0$ and $\nabla_{s_i}\gamma_i = 0$ at the destination points, we have $\nabla_{s_i}\varphi_i|_{s\in F_{id}} = 0$ in view of (9.3). Therefore, the destination points $s \in F_{id}$ are also the critical points of φ_i. Then, the Hessian of φ_i evaluated at the destination points is calculated as

$$\nabla_{s_i}^2\varphi_i|_{s\in F_{id}} = \frac{1}{\beta_i^{\frac{1}{\alpha}}}\nabla_{s_i}^2\gamma_i|_{s\in F_{id}}$$
$$= \frac{1}{\beta_i^{\frac{1}{\alpha}}}D_{s_i}(2\nabla_{s_i}^2 H\nabla_{s_i}H)|_{s\in F_{id}}$$
$$= \frac{2}{\beta_i^{\frac{1}{\alpha}}}(\nabla_{s_i}^2 H)^T\nabla_{s_i}^2 H|_{s\in F_{id}}$$

It follows from the assumption and Lemma 9.1 that $\nabla_{s_i}^2\varphi_i|_{s\in F_{id}} \succ 0$, which implies the destination points $s \in F_{id}$ correspond to the minimums of φ_i. □

Lemma 9.7 *For the agent with dynamics (9.1) and control law (9.2), the trajectory of the system converges to the set*

$$\{s : \|\nabla_{s_i}\varphi_i\| \leq \bar{\epsilon}, \forall i \in E_n\}$$

in finite time if $\alpha > \theta(\bar{\epsilon})$, *where the scalar* $\theta(\bar{\epsilon}) > 0$ *is related to* $\bar{\epsilon} > 0$.

For the proof, the reader is referred to Proposition 2 in [18].

Clearly, Lemma 9.2 implies that the trajectories of agents do not converge to the set of undesired points. Moreover, with Lemmas 9.3, 9.4 and 9.5, we see that there are no attractors in the region $F_0(\epsilon) \bigcup F_1(\epsilon) \bigcup F_2(\epsilon) \bigcup F_3(\epsilon)$ and on the boundary $\partial \mathscr{W}_i$. Finally, Lemmas 9.6 and 9.7 show that the agents will approach to the destination points. Therefore, we get the following conclusion easily.

Theorem 9.2 *Suppose that the two assumptions hold. If the initial positions of agents are not located at the undesired points and* $\kappa > \alpha > \max\{\eta(\epsilon), \theta(\bar{\epsilon})\}$ *with* $\epsilon < \epsilon_1$, *then the trajectories of agents with dynamics (9.1) and control law (9.2) almost converge to arbitrarily small neighborhood of the destination points.*

Remark 9.7 According to the 2nd assumption and the definition of F_{id}, the destination points actually correspond to the local maximums of $H(s)$.

Remark 9.8 Generally speaking, it is difficult for multiple agents to optimize the coverage index in the non-convex region using traditional path planning algorithms. With navigation functions, the coverage path planning of agents in the region with obstacles can be formulated as the optimization problem with constraints. Moreover, the regions, which is topologically equivalent to the sphere through diffeomorphisms, can also be dealt with.

9.4 Numerical Simulation

This section presents simulation results of coverage control algorithm with the agent dynamics in Theorem 9.1. Six agents are employed to cover an uncertain rectangular region, and the event density function is given by

$$\Phi(q) = 4 - 0.1\|q - q_0\|$$

with $q_0 = [0, 20]$. The function $f(s, q)$ is described by

$$f(s, q) = 1 - \prod_{i=1}^{n}(1 - e^{-\lambda\|s_i - q\|}),$$

where s_i denotes the position of the i-th agent and λ refers to a positive parameter with $\lambda = 1$. The sensing radius of agents is 2.5 m. The simple Euler method is adopted to implement the coverage algorithm with sampling period $T = 0.01$s. Figure 9.2 illustrates the deployment of six agents at four different time steps (i.e., 1, 100, 200, 1000). Initially, the agents stay on a straight line. Then all the agents move towards the center of coverage region as the event density is relatively large, as demonstrated by the color map. Finally, the agents converge to a flower-like structure at about

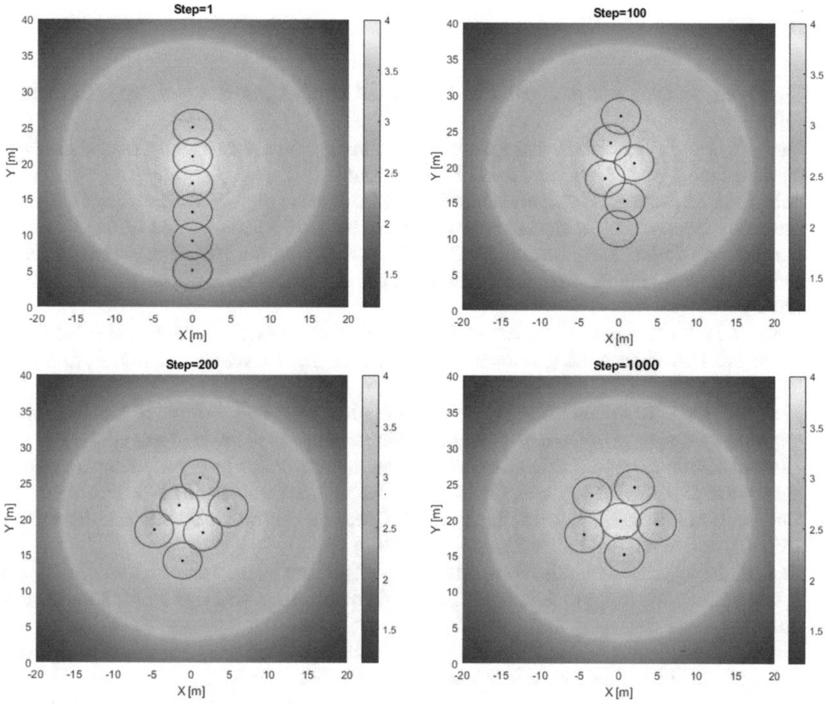

Fig. 9.2 Implementation of distributed coverage control algorithm in uncertain region

$t = 10$s, which allows to maximize the detection probability of random events in the given coverage region.

9.5 Conclusions

In this chapter, we consider the coverage problem of multiple mobile agents in the given region. A decentralized cooperative controller is developed to maximize the joint effectiveness of actions imposed by multiple agents. Moreover, a decentralized navigation function is designed to drive the multiple agents towards the destination points while avoiding the obstacles and undesired points, and the local maximum of the objective function can be achieved. Further research may include investigating the coverage of multiple agents with nonholonomic dynamics and relaxing the two assumptions.

References

1. Shamman, J.: Cooperative Control of Distributed Multi-Agent Systems. Wiley, Chichester (2007)
2. Hong, Y., Hu, J., Gao, L.: Tracking control for multi-agent consensus with an active leader and variable topology. Automatica **42**, 1177–1182 (2006)
3. Schwager, M., Julian, B., Angermann, M., Rus, D.: Eyes in the sky: decentralized control for the deployment of robotic camera networks. Proc. IEEE **99**(9), 1541–1561 (2011)
4. Howard, A., Parker, L.E., Sukhatme, G.S.: Experiments with a large heterogeneous mobile robot team: exploration, mapping, deployment and detection. Int. J. Robot. Res. **25**, 431–447 (2006)
5. Casbeer, D.W., Kingston, D.B., Beard, R.W., Mclain, T.W., Li, S.M., Mehra, R.: Cooperative forest fire surveillance using a team of small unmanned air vehicles. Int. J. Syst. Sci. **37**, 351–360 (2006)
6. Zhai, C., Hong, Y.: Decentralized algorithm for online workload partition of multi-agent systems. In: Proceedings of Chinese Control Conference, pp. 4920–4925. Yantai (2011)
7. Hussein, I., Stipanović, D.: Effective coverage control for mobile sensor networks with guarteed collision avoidance. IEEE Trans. Control Syst. Technol. **15**(4), 642–657 (2007)
8. Cortés, J., Martínez, S., Karatas, T., Bullo, F.: Coverage control for mobile sensing network. IEEE Trans. Robot. Autom. **20**(2), 243–255 (2004)
9. Zhai, C., Hong, Y.: Decentralized sweep coverage algorithm for uncertain region of multi-agent systems. In: Proceedings of American Control Conference, pp. 4522–4527. Montréal, Canada (2012)
10. Wagner, I., Lindenbaum, M., Bruckstein, A.: Distributed covering by ant-robots using evaporating traces. IEEE Trans. Robot. Autom. **15**(5), 918–933 (1999)
11. Rekleitis, I., New, A.P., Rankin, E.S., Choset, H.: Efficient boustrophedon multi-robot coverage: an algorithmic approach. Ann. Math. Artif. Intell. **52**, 109–142 (2008)
12. Cassandras, C., Li, W.: Sensor networks and cooperative control. Eur. J. Control **11**(4–5), 436–463 (2005)
13. Koditschek, D.E., Rimon, E.: Robot navigation functions on manifolds with boundary. Adv. Appl. Math. **11**, 412–442 (1990)
14. Ghaffarkhah, A., Mostofi, Y.: Communication-aware target tracking using navigation functions-Centralized case. In: Proceedings of International Conference Robotics Communication, pp. 1–8. Odense (2009)
15. Carmela, M., Gennaro, D., Jadbabaie, A.: Formation control for a cooperative multi-agent system using decentralized navigation functions. In: Proceedings of American Control Conference, pp. 1346–1351. Minneapolis, Minnesota (2006)
16. Kan, Z., Dani, A.P., Shea, J.M., Dixon, W.E.: Network connectivity preserving formation stabilization and obstacle avoidance via a decentralized controller. IEEE Trans. Autom. Contr. **57**(7), 1827–1832 (2012)
17. Rimon, E., Koditschek, D.E.: Exact robot navigation using artificial potential functions. IEEE Trans. Robot. Autom. **8**(5), 501–518 (1992)
18. Dimarogonas, D., Frazzoli, E.: Analysis of decentralized potential field based multi-agent navigation via primal-dual Lyapunov theory. In: Proceedings of the IEEE Conference Decision and Control, pp. 1215–1220. Atlanta (2010)

Chapter 10
Summary and Future Work

10.1 Summary

This book focuses on the coverage control problem of MAS in uncertain environments and proposes several novel control approaches to dealing with various cooperative coverage missions. In particular, a theoretical formulation of multi-agent sweep coverage is developed to complete the workload on the coverage region. In order to minimize the coverage time, a divide-and-conquer scheme is proposed to partition the whole region into multiple sub-regions, and each agent is only required to complete the workload in its own sub-region. Meanwhile, MAS implements the distributed algorithm to partition the unexplored regions for sweep coverage in the future. The above scheme enables MAS to cope with the unknown environment and complete the sweeping mission as soon as possible. To assess the proposed sweeping approach, the time error between the actual coverage time and the optimal time is estimated and its upper bound is derived as well. On the basis of the proposed sweep coverage formulation, adaptive sweep coverage algorithm is developed to allow for the unknown workload in the coverage region. In addition, a fully distributed control algorithm is designed to sweep a certain region by capitalizing on workload memory. Besides multi-agent sweep coverage, the cooperative blanket coverage of MAS is also investigated and applied to the interception of supersonic flight vehicles. A coverage-based guidance algorithm is developed to maximize the joint interception probability of interceptors of low maneuverability against supersonic flight vehicles equipped with decoys. Moreover, the blanket coverage approach can be employed to deploy WSN for the environment monitoring (e.g., geo-hazards monitoring and early warning). Further, a coverage-based routing algorithm for UGVs is proposed to choose the optimal path for arriving at the destination as soon as possible while alleviating the congestion level of transportation system.

C. Zhai et al., *Cooperative Coverage Control of Multi-Agent Systems and its Applications*,
Studies in Systems, Decision and Control 408,
https://doi.org/10.1007/978-981-16-7625-3_10

10.2 Future Work

- **Wide-Area Cooperative Coverage Control via UAV-Satellite Coordination**
The wide-area cooperative coverage of UAVs normally calls for the monitoring and reconstruction of ever changing environments, which resorts to the satellite. The surface displacement can be detected by the radar on the satellite using remote sensing technologies. It is of great significance to investigate the distributed coverage control algorithm of UAV for the detection of concerned random events in the dynamical environments with the assistance of surface displacement information from the satellite. Figure 10.1 illustrates the UAV-satellite coordination for the cooperative coverage of wide-area uncertain environments.
- **Coverage Control Design for Time-Varying Irregular Regions**
In some circumstances, the boundary of coverage region is time varying, which inevitably brings uncertainties and challenges to the coverage control of MAS. This will give rise to the redeployment of MAS for the optimization of objective functions. If the coverage region becomes non-convex, it significantly affects the optimal solution of coverage control problem and could cause the surge of computational complexity. Usually, sub-optimal solutions are available for numerical algorithms of multi-agent coverage control. For example, Fig. 10.2 describes the evolution of coverage region, which leads to the redeployment of MAS.
- **Mathematical Framework for Coverage-Based Stochastic Routing Strategy**
Chapter 9 presents a type of deterministic coverage-based routing algorithm for relieving the traffic congestion during rush hours. Nevertheless, there is no mathematical frameworks to formulate the routing problem of UGVs in dynamic uncertain environments, which results in the failure of theoretical assessment and analysis of the proposed routing algorithm. Numerical simulations are not convincing

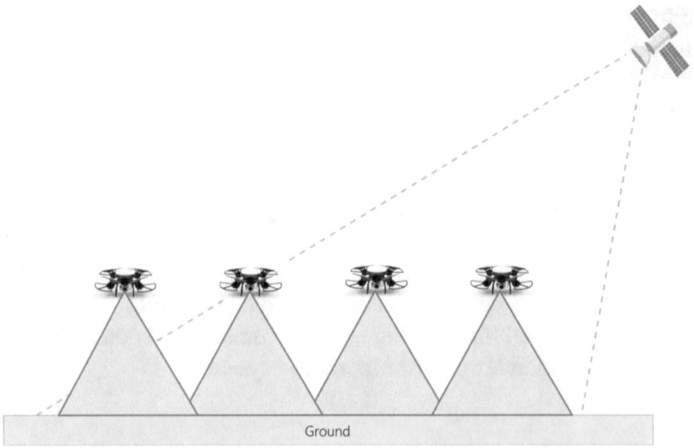

Fig. 10.1 Coordination between UAVs and satellite for wide-area coverage

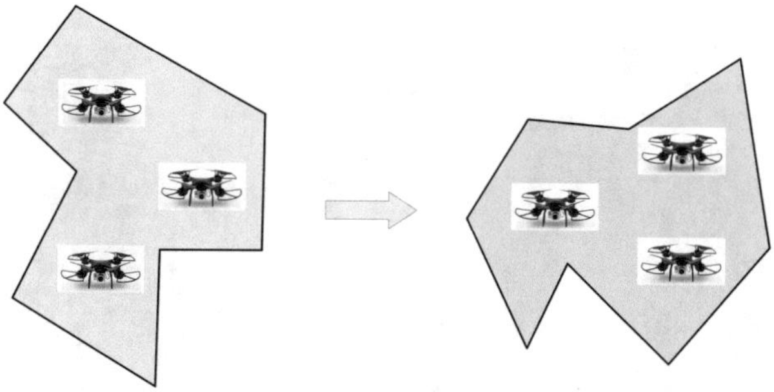

Fig. 10.2 Cooperative coverage of UAVs in time-varying irregular regions

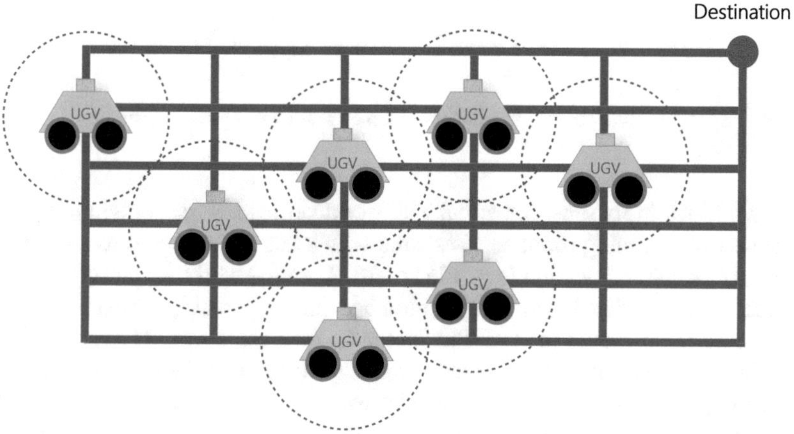

Fig. 10.3 Cooperative coverage of UAVs in time-varying irregular regions

in terms of demonstrating the advantages of routing algorithms, as compared to other existing algorithms. Therefore, it is necessary to come up with a mathematical framework for the UGV routing problem from a stochastic perspective. Figure 10.3 gives a brief idea on the design of stochastic routing algorithm, where each UGV collects the data on congestion level in its sensing region (i.e., red circle) at the intersection and decides to choose the optimal path with a given probability. This probability depends on the congestion level of relevant paths as well as the distance to the destination (i.e., red ball).

- **Experimental Validation of Cooperative Coverage Control Algorithm**
  Since the experimental setup is not physically available for the time being, experimental validations are not provided in this book. In the next step, we plan to implement the proposed coverage control algorithms using robots or UAVs. The schematic diagram of experimental setup of the multi-robot sweep coverage is shown in Fig. 10.4. For simplicity, three sweeping robots are employed to complete the dust on the floor. Each sweeping robot is equipped with a sensor to detect

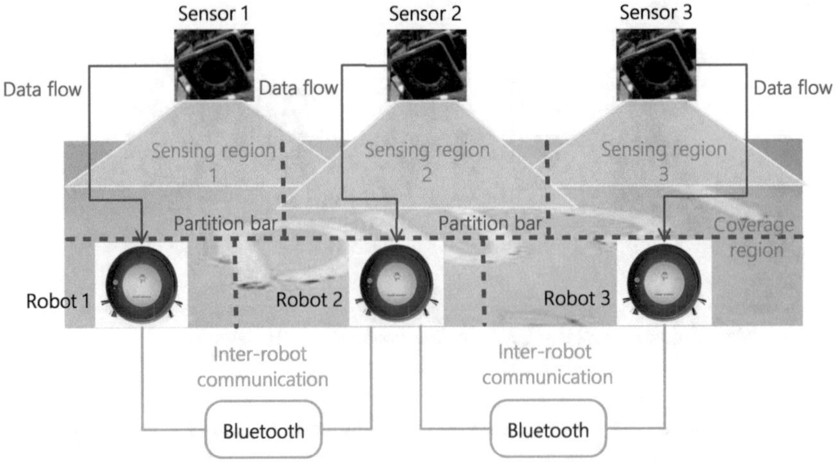

Fig. 10.4 Schematic diagram of experimental setup on multi-robot sweep coverage

the workload in its sensing region. The workload data is transmitted to the storage unit of sweeping robot and is shared with adjacent sweeping robots through inter-robot communication (e.g., bluetooth). Then the workload partition is taken in real time to update the virtual partition bars on the next stripe. Meanwhile, each sweeping robot moves to clean the dust in its own sub-stripe. The main difficulties in implementing the sweep coverage algorithm lie in the real-time detection and computation of workload on sub-stripes, which requires the effective integration of high precision sensors, image processing module and positioning system. Experiments of multi-robot sweep coverage will be the focus of our future work.

Appendix
Mathematical Concepts

In this appendix, some key mathematical definitions are provided to facilitate theoretical analysis of cooperative coverage problem.

A.1 Algebraic Graph Theory

A directed graph $\mathcal{G} = (\mathcal{V}, \mathcal{E})$ is composed of node set $V = \{1, 2, ..., n\}$ and arc set $\mathcal{E} \subseteq \mathcal{V} \times \mathcal{V}$. Let $A = [a_{ij}]_{n \times n}$ denote an adjacency matrix with non-negative elements a_{ij}, which is positive if and only if $(i, j) \in \mathcal{E}$. In addition, $\mathcal{N}_i = \{j \in \mathcal{V} | (i, j) \in \mathcal{E}\}$ represents the set of neighbors for node i. Figure A.1 presents an example of unweighted graph with 4 nodes and 4 edges, and its adjacency matrix is given by

$$A = \begin{pmatrix} 0 & 1 & 1 & 0 \\ 1 & 0 & 1 & 0 \\ 1 & 1 & 0 & 1 \\ 0 & 0 & 1 & 0 \end{pmatrix}$$

To investigate the graph connectivity, Laplacian matrix is defined as follows [1].

Definition A.1 A matrix $\mathcal{L} \in R^{n \times n}$ is Laplacian if $\mathcal{L}1_n = 0_n$ and its off-diagonal entries are non-positive.

The Laplacian matrix of a weighted digraph can be obtained by

$$\mathcal{L} = \texttt{diag}(A1_n) - A$$

If $\mathcal{L}$ is symmetric, the eigenvalues of $\mathcal{L}$ can be ordered as

$$0 = \lambda_1 \leq \lambda_2 \leq \cdots \leq \lambda_n$$

C. Zhai et al., *Cooperative Coverage Control of Multi-Agent Systems and its Applications*, Studies in Systems, Decision and Control 408,
https://doi.org/10.1007/978-981-16-7625-3

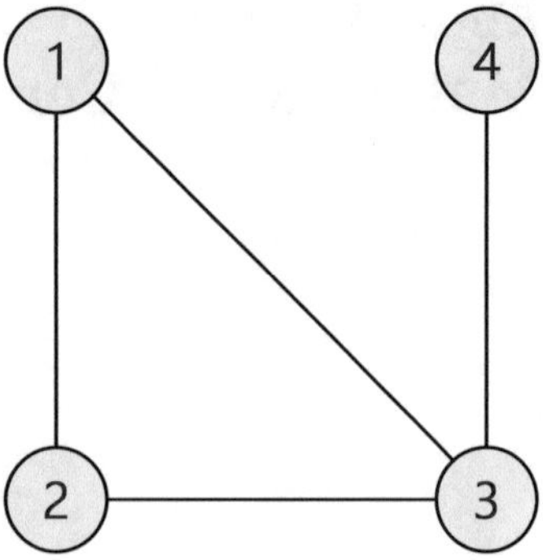

Fig. A.1 An example of unweighted graph

and the graph $\mathcal{G}$ is connected if and only if $\lambda_2 > 0$. Thus, the smallest nonzero eigenvalue λ_2 is called the algebraic connectivity of $\mathcal{G}$.

A.2 Input-to-State Stability

Before introducing input-to-state stability, it is necessary to define two special comparison functions (i.e., class $\mathcal{K}$ and class $\mathcal{KL}$ functions).

Definition A.2 A continuous function $\alpha : [0, a) \to [0, \infty)$ is said to belong to class $\mathcal{K}$ if it is strictly increasing and $\alpha(0) = 0$. It is said to belong to class $\mathcal{K}_\infty$ if $a = \infty$ and $\alpha(r) \to \infty$ as $r \to \infty$.

Definition A.3 A continuous function $\beta : [0, a) \times [0, \infty) \to [0, \infty)$ is said to belong to class $\mathcal{KL}$ if, for each fixed s, the mapping $\beta(r, s)$ belongs to class $\mathcal{K}$ with respect to r and, for each fixed r, the mapping $\beta(r, s)$ is decreasing with respect to s and $\beta(r, s) \to 0$ as $s \to \infty$.

Consider a dynamical system

$$\dot{x} = f(t, x, u), \tag{A.1}$$

where the function $f : [0, \infty) \times R^n \times R^m \to R^n$ is piecewise continuous in t and locally Lipschitz in x and u. In addition, the input $u(t)$ is a piecewise continuous, bounded function with respect to t for all $t \geq 0$. It is assumed that the unforced system

$$\dot{x} = f(t, x, 0) \tag{A.2}$$

has a globally uniformly asymptotically stable equilibrium at the origin $x = 0$.

Now, we present the formal definition of input-to-state stability as follows [2].

Definition A.4 The dynamical system (A.1) is input-to-state stable if there exist a class $\mathcal{KL}$ function β and a class $\mathcal{K}$ function γ such that for any initial state $x(t_0)$ and any bounded input $u(t)$, the solution $x(t)$ exists for all $t \geq t_0$ and satisfies

$$\|x(t)\| \leq \beta(\|x(t_0)\|, t - t_0) + \gamma\left(\sup_{t_0 \leq \tau \leq t} \|u(\tau)\|\right).$$

References

1. Dorfler, F., Simpson-Porco, J., Bullo, F.: Electrical networks and algebraic graph theory: models, properties, and applications. Proc. IEEE **106**(5), 977–1005 (2018)
2. Khalil, H., Grizzle, J.: Nonlinear Systems, vol. 3. Prentice Hall, Upper Saddle River (2002)

Printed in the United States
by Baker & Taylor Publisher Services